Operations Management and Management Science

Edited by Fausto Pedro García Márquez

Published in London, United Kingdom

Operations Management and Management Science
http://dx.doi.org/10.5772/intechopen.96833
Edited by Fausto Pedro García Márquez

Contributors
Roberto Sarmiento, Kotomichi Matsuno, Jiahua Weng, Noriyuki Hosokawa, Takahiro Ohno, Nneoma Benita Amos, Edafe Bawa Dogo, John Kolade Obamiro, Wasiu Oluwasegun Kehinde, Jonathan Ehimen Ekpudu, Oluwasegun David, Yu Zhao, Yiannis Laouris, Alexsandro Silva Solon, Marcos Paulo Leite Silveira, Marcos J. Negreiros, Nelson Maculan, Augusto W.C. Palhano, Albert E.F. Muritiba, Pablo L.F. Batista, Ashok G. Verghese, Thalavaipillai Sudalaimuthu, Pon Ramalingam, Călin Corciovă, Robert Fuior, Doru Andriţoi, Cătălina Luca

First published in London, United Kingdom, 2023 by IntechOpen
IntechOpen is the global imprint of INTECHOPEN LIMITED, registered in England and Wales, registration number: 11086078, 5 Princes Gate Court, London, SW7 2QJ, United Kingdom

British Library Cataloguing-in-Publication Data
A catalogue record for this book is available from the British Library

Additional hard and PDF copies can be obtained from orders@intechopen.com

Operations Management and Management Science
Edited by Fausto Pedro García Márquez
p. cm.
Print ISBN 978-1-80355-982-7
Online ISBN 978-1-80355-981-0
eBook (PDF) ISBN 978-1-80355-983-4

For EU product safety concerns:
IN TECH d.o.o., Prolaz Marije Krucifikse Kozulić 3, 51000 Rijeka, Croatia,
info@intechopen.com or visit our website at intechopen.com.

Meet the editor

Fausto Pedro García Márquez has been a Full Professor at the University of Castilla-La Mancha (UCLM), Spain, since 2013, is an Honorary Senior Research Fellow at Birmingham University, UK and a lecturer at the Postgraduate European Institute. He was a Senior Manager at Accenture from 2013 to 2014 and is currently the director of Ingenium Research Group obtained his European Ph.D. with maximum distinction. Since 2021 he has been a Senior Member of IEEE and an Honoured Honorary Member of the Research Council of the Indian Institute of Finance. He became the Committee Chair of the International Society for Management Science and Engineering Management (ISMSEM) in 2020. Among his awards are the Nominate Prize (2022), Grand Prize (2021), Runner Prize (2020) and Advancement Prize (2018), Runner (2015), Advancement (2013) and Silver (2012) from the International Society of Management Science and Engineering Management (ICMSEM), and the First International Business Ideas Competition 2017 Award (2017). He has published more than 200 papers and is the author or editor of more than 45 books, over 100 international chapters, and six patents. He has been the principal investigator in four European projects, eight national projects, and more than 150 projects for universities and companies. His main interests are: artificial intelligence, maintenance, management, renewable energy, transport, advanced analytics, and data science.

Contents

Preface

Operations management may be defined as the efficient transformation of inputs to outputs, carried out according to customer needs and considering the limitations present in the process. It, therefore, involves process optimization, from design to future requirements, the management of materials and products, production, and such other areas as maintenance and quality control. The products may be considered goods or services, while the inputs may be human resources, financial support, information, materials, energy, and so on.

Industry uses a large amount of data which, with appropriate analytics, provides more information and leads to an increase in the efficiency of operations management. In the various disciplines such as forecasting, capacity planning, location, layout, and integration of activities, new methods are constantly evolving to solve problems related to operations management, for example, dynamic analysis, computational techniques, probabilistic methods, and mathematical optimization techniques. Most of these new methods are based on artificial intelligence, for example, machine learning, deep learning, natural language processing, expert systems, data mining, support vector machines, and so on. Heuristics and metaheuristics algorithms are often found in optimization problems, for example, genetic algorithms, colony optimization, cuckoo search optimization, greedy randomized adaptive search procedure, particle swarm optimization, etc. Hybrid algorithms that combine several methods are being used today to improve solutions or to solve complex and robust problems that cannot be solved in another way.

This book introduces the main concepts and theoretical frameworks in operations management, as well as the latest empirical research findings from around the world. It also presents case studies from several industries, in which standard and novel algorithms are used to solve problems. Finally, future applications, trends and future work are suggested to improve the solutions presented.

Fausto Pedro García Márquez
Ingenium Research Group,
University of Castilla-La Mancha,
Ciudad Real, Spain

Chapter 1

Differences between Universal-Deterministic and Probabilistic Hypotheses in Operations Management Research

Roberto Sarmiento

Abstract

Very few papers in the operations management (OM) field have taken the themes of universal-deterministic (UD) and probabilistic hypotheses as their main topics of investigation and discussion. Our investigation continues a recent line of research that focuses on a better understanding of these critical issues. Specifically, we attempt to respond to some pointed criticisms that experts in the field have made when the topic UD and probabilistic hypotheses have emerged in academic settings/discussions. A detailed analysis of those criticisms shows that they lack merit, thereby reinforcing our argument that it is most important to distinguish between the two types of scientific hypotheses in order to advance in the rigor of OM theoretical and empirical research. Ideas for future research are outlined.

Keywords: universal-deterministic hypotheses, probabilistic hypotheses, case study research, quantitative studies, qualitative studies

1. Introduction

Previous investigations (e.g., [1, 2]) have attempted to provide a better understanding and awareness vis-à-vis the two types of hypotheses that exist in all fields of empirical science: universal-deterministic (UD) and probabilistic. In particular, those papers emphasize the potential problems of not acknowledging the theoretical/empirical differences between these two types of scientific propositions. Continuing with this line of research, our paper deals with some objections that have been made by experts in the operations management (OM) field[1]. More specifically, our paper discusses why those criticisms are not justified. This article strengthens the argument about the importance of acknowledging the distinct characteristics that UD and probabilistic hypotheses possess. This topic is important, not only in operations management research, but in all areas of empirical science.

In sections 2, 3, and 4, we present the aforementioned criticisms and explain why those objections lack justification. In Section 5, we further explore (based on Popperian logic and methodology) how falsified hypotheses can still be of practical use. Section 6 offers some concluding remarks.

2. Criticism (a): "measurement variability messes up deterministic hypotheses. This problem turns a deterministic relationship into a probabilistic one"

There is no question that variability in the measurement of individual observations is an important issue in the empirical sciences. Nevertheless, it is not accurate to say that measurement variability transforms a UD hypothesis into a probabilistic one. These are two separate issues: 1) the problems with the measurement of empirical evidence (e.g., individual observations) when testing a hypothesis of interest, and 2) the logical form and implications of UD hypotheses.

Popper (e.g., [7], p. 63) acknowledges the potential problems that exist with respect to measurement variability. His recommendation is that there should be "methodological rules" by which decisions regarding empirical evidence in support of or against a given hypothesis may be agreed upon: "... Inter-subjectively testable experiments are either to be accepted, or to be rejected in the light of counter-experiments." Or as Dienes ([8], p. 20, his italics) puts it,

[1] Ross [3] says "[I]t is often stated that, unlike classical physics, Quantum Physics is not deterministic. This statement is not really correct ...". With this assertion, Ross begins a detailed explanation of a concept that he calls "probabilistic determinism." Later in his treatise, Ross writes that "[T]his means we do indeed have determinism, but only determinism of probability distributions of positions and momentum, as opposed to determinism of their exact values...". "Thus, although Quantum Physics does not allow us to determine where a particular photon will land, it does allow us to determine where we will find dense and sparse regions – and in this sense it is deterministic," Ross concludes. To reinforce his point, Ross explains that "probabilistic determination" is a viewpoint/approach that is tacitly/implicitly accepted/used in other disciplines, such as the Social Sciences. In the process of publishing OM papers that deal with the topics of UD and probabilistic hypotheses, we have received criticisms from anonymous referees that are similar to Ross' arguments, e.g., "the probabilistic versus deterministic distinction is basically a red herring," "something that is impossible according to a deterministic hypothesis is basically the same as something that has a one-in-a-billion chance of happening according to a probabilistic hypothesis." Referees are supposed to have considerable expertise on the subjects that are presented in the investigations they evaluate. Therefore, it is reasonable to conclude that if expert reviewers are not aware of the differences between UD and probabilistic hypotheses, then it is possible that the rest of the field is also unaware of these themes. We also note that highly influential works that offer guidance about case study research methodology (e.g., [4–6]) do not offer a detailed discussion of the differences between UD and probabilistic hypotheses. Moreover, when these topics have been brought up in informal gatherings and discussions with colleagues in other areas of business management research, a lack of awareness/understanding of these types of scientific propositions has also been identified. We argue that these situations justify our investigation. We are quoting or paraphrasing the objections that are dealt with in the paper.

> According to Popper, observation statements are finally accepted only by decision or agreement. Finally, there comes a point where everyone concerned feels the observation statement is sufficiently motivated that no one wishes to deny it. Considerable work may be needed to reach that point; and even then the decision to accept an observation may be overturned by new considerations. [...] In the end we must decide which observation statements we will accept. The decision is fallible and amounts to tentatively accepting a low-level empirical hypothesis which describes the effect: For example, accepting an observation statement amounts to accepting a hypothesis that "Peter is an extrovert", or "This extrovert was asleep at 7 am" and so on.

To illustrate Popper's point, we recall the story of the experiments and events that corroborated Einstein's general theory of relativity. According to Folsing, ([9], p. 443),

> The numerical values obtained were 1.98 ± 0.12 seconds of arc for the Sobral pictures and 1.61 ± 0.30 seconds of arc for the pictures from Principe (which had been impaired by clouds). Both results ruled out the "Newtonian" value; their mean value was almost exactly equal to Einstein's prediction of 1.74 seconds of arc.

This famous episode in the history of science shows that irrespective of the variability/errors that sometimes are inevitable when measuring empirical evidence, scientists can still develop methodological rules/agreements in order to judge whether the predictions/implications of a UD hypothesis have been corroborated or refuted. Put differently, the variability in the measurement of empirical evidence does not turn a UD hypothesis (e.g., Einstein's) into a probabilistic one.

We now offer a more recent example. Overbye [10] reported that in September of 2011, a group of scientists announced that they had measured a batch of subatomic particles (neutrinos) traveling faster than the speed of light. This appeared to falsify the UD law that negates that specific occurrence ("no object can travel faster than the speed of light"). However, Overbye then quoted a scientist as saying that "[T]he evidence is beginning to point toward the (results showing falsifying evidence of the UD law) being an artifact of the measurement" (see also [11] for more reports about the same episode).

The neutrinos story was very interesting to follow because the whole process – from the initial claims of falsification to the conclusion that errors had been made during the experiments – was consistent with the logic and methodology that Popper proposed in order to test a UD law. It also serves to illustrate once more that regardless of the different potential difficulties that can be present when testing a UD hypothesis, scientists can still come to the conclusion (based on methodological rules/conventions) that a given UD proposition has been refuted or corroborated. This helps us to affirm that the different difficulties that might appear when testing a UD hypothesis do not turn it into a probabilistic one.

Response to criticism (a): potential measurement problems that exist when testing the implications of UD hypotheses do not turn them into probabilistic ones. Instead, scientists can/should arrive at methodological rules/agreements upon which decisions should be made about empirical evidence that appear to corroborate or refute a hypothesis of interest.

3. Criticism (b): "the probabilistic versus deterministic distinction is basically a red herring," "something that is impossible according to a deterministic hypothesis is basically the same as something that has a one-in-a-billion chance of happening according to a probabilistic hypothesis"

These criticisms arguably reflect the prevalent thinking among scientists that are familiar with the frequentist approach to probability and statistical hypothesis testing[2]. For the sake of argument, we will assume that this approach is an adequate way to apply Popperian falsificationism when testing a probabilistic hypothesis[3]. In this approach, researchers usually have to formulate a null hypothesis (Ho) and an alternative hypothesis (H1). We will use these assumptions and conventions to explain one of the theoretical/empirical differences between UD and probabilistic hypotheses.

We first discuss a situation where researchers are interested in testing whether a variable "A" (the cause) has an effect on variable "B". We model this relationship using the usual principles and assumptions of standard linear regression analysis. We state Ho as a general proposition: "A has no noticeable effect on B" (e.g., Ho: $\beta 1 = 0$). We also establish H1 as "A has a noticeable effect (e.g., "statistically significant") on B" (e.g., $\beta 1 \neq 0$). Let us also suppose that two different investigations are performed: one to test a probabilistic relationship between A and B, and the other one to test its UD version. To be clear that we are dealing with two different and separate situations, we will rename the variables as *Ap/Bp* in the case of a probabilistic relationship and *Ad/Bd* to model its UD version. We can now restate both H1's as "*Ap* is likely to have a positive and significant effect on *Bp*," and "the effect of *Ad* on *Bd* is positive and linear," respectively. **Figures 1** and **2** illustrate the results of the hypothetical investigations of the probabilistic and UD relationships:

These figures show evidence supporting the alternative hypothesis (e.g., H1: "A has a noticeable and positive effect on B") in both its probabilistic and UD forms. Based on this, the assertion that there are no practical differences between these two types of hypotheses would appear to be true. Put differently, since **Figures 1** and **2** support the idea that Ho ("A has no noticeable effect on B") in its probabilistic and UD forms has been "practically falsified" and "falsified," respectively, it does seem as if arguing that UD and probabilistic hypotheses are different in nature is indeed a red herring.

We now proceed to explain that, in spite of what was discussed previously, there are clear theoretical/empirical differences between UD and probabilistic hypotheses. To accomplish our objective, we again model a causal relationship between two different variables: C ("the cause") and D ("the effect"). We make the same assumptions as before, with only one difference: we now suppose that there is prior corroborating evidence showing a noticeable and positive effect of C on D. We state the probabilistic version of this relationship as "*Cp* is likely to have a positive and noticeable effect on *Dp*" (i.e., Ho: $\beta 1 \neq 0$). The UD form of this relationship could be phrased along the lines of "*Cd* has a linear and positive effect on *Dd*" (e.g., Ho $\beta 1 = 1$). This is a different situation to the one that was analyzed in the previous paragraphs (Ho: $\beta 1 = 0$). Let

[2] For a more technical discussion on those topics, we recommend Popper [7] and Miller [12].

[3] Popper ([7], p. 183n) affirms that due to their logical form, probabilistic hypotheses are not directly falsifiable. However, he also says that scientists can arrive at "...the adoption of a methodological rule..., which makes probability hypotheses falsifiable." In this way, probabilistic hypotheses can attain a scientific status, because once a methodological rule has been adopted, their implications can be tested empirically (i.e., their predictions can be contradicted by empirical evidence).

DOI: http://dx.doi.org/10.5772/intechopen.1000218

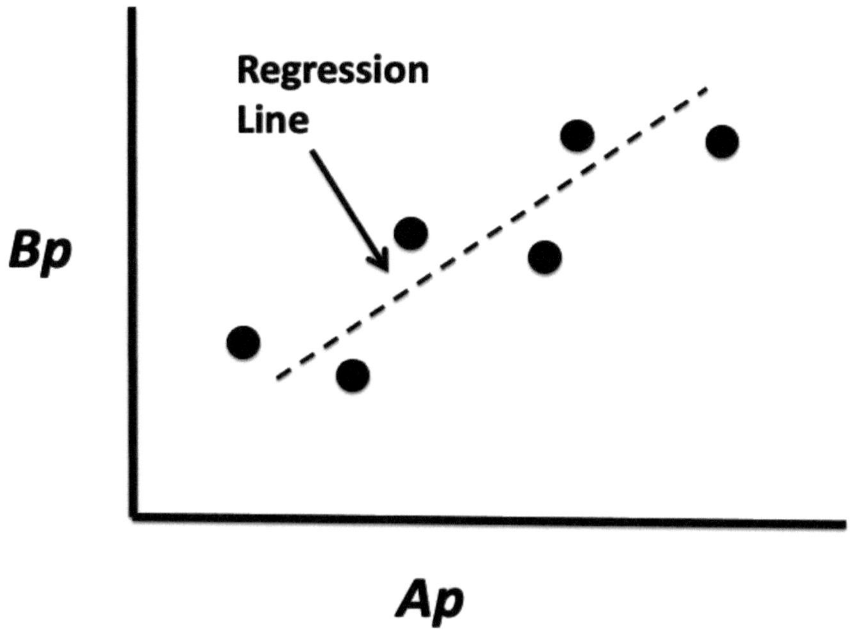

Figure 1.
Evidence supporting a probabilistic relationship between Ap and Bp.

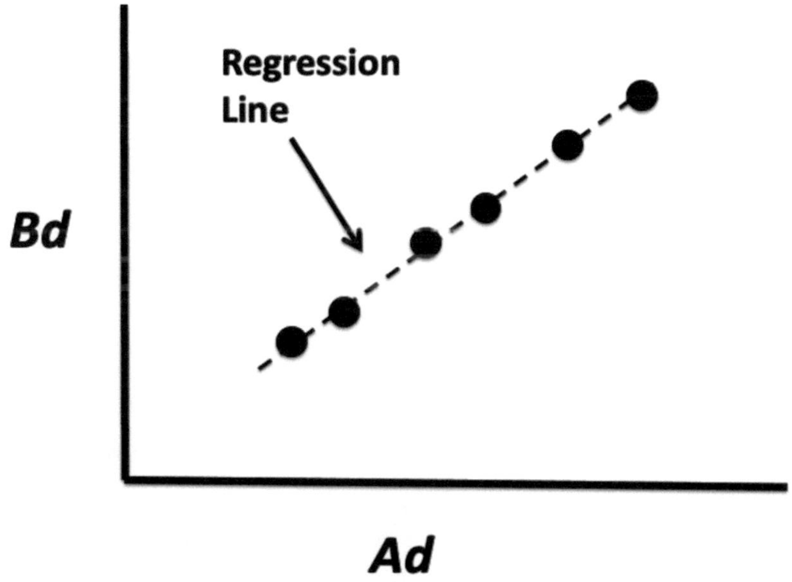

Figure 2.
Evidence supporting a UD relationship between Ad and Bd.

us also suppose that two investigations are run to examine these two relationships. **Figures 3** and **4** show the results of these hypothetical investigations.

These illustrations show one of the theoretical/empirical differences between UD and probabilistic hypotheses. **Figure 3** shows evidence that the probabilistic relationship

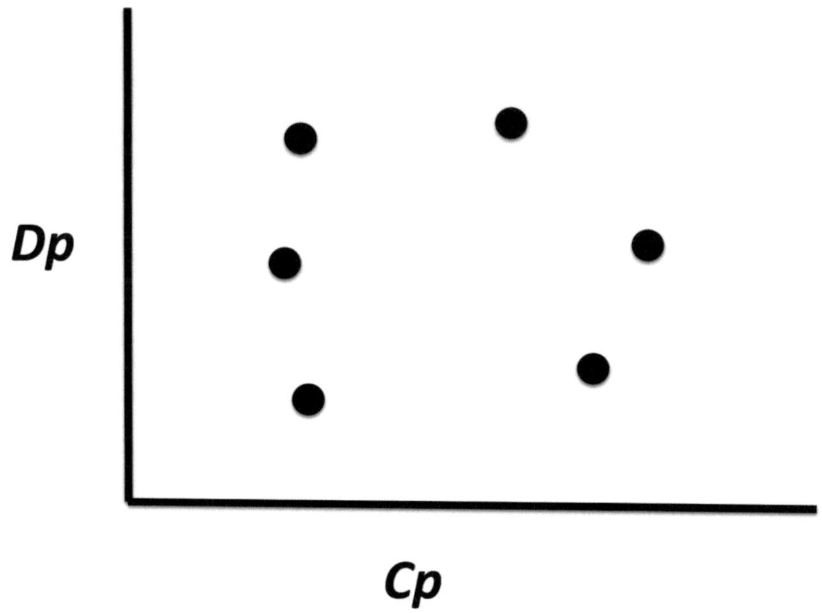

Figure 3.
Evidence that "practically falsifies" a probabilistic relationship between Cp and Dp.

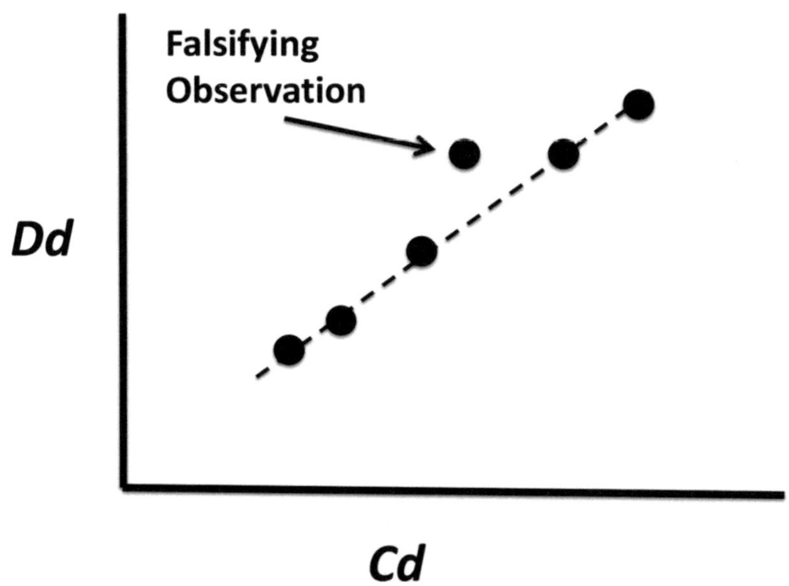

Figure 4.
Evidence that falsifies (logically and strictly) a UD relationship between Cd and Dd.

between *Cp* and *Dp* has been "practically falsified." Likewise, **Figure 4** shows evidence that the UD relationship between *Cd* and *Dd* has been falsified (logically and strictly). In short, Ho for both the probabilistic and UD forms of the causal relationship between C and D has been "practically falsified" and "falsified," respectively.

While it is true that in both cases there appears to be undermining/falsifying evidence against Ho *(when Ho establishes a noticeable and positive relationship between the variables)*, it is obvious that in theoretical/empirical terms, researchers finding evidence that "practically falsifies" a probabilistic hypothesis cannot continue to affirm that there is a noticeable and positive effect of *Cp* on *Dp*. **Figure 3** shows evidence of this non-relationship. Therefore, for all practical and theoretical purposes, researchers should be inclined to conclude that there is no evidence of a relationship between *Cp* and *Dp*. On the other hand, researchers who find falsifying evidence of the UD relationship between *Cd* and *Dd* should be inclined to say that even though the UD relationship has been (logically and strictly) falsified, there still appears to be a noticeable and positive relationship between *Cd* and *Dd*. In other words, the relationship between these two variables, for all practical and theoretical purposes, can still be considered as positive and noticeable. In short, the falsified UD relationship can now be stated as a probabilistic one: *Cd* is likely to have a noticeable and positive effect on *Dd*.

To complement the current discussion, we now offer an example of a relationship where the dependent and independent variables can be present or absent. To avoid repetition, now we will only model the situation where Ho states a prior noticeable effect of E (the cause) on F (the effect). The probabilistic form of this relationship can be phrased along the lines of "if E occurs/is present, then F is more likely to occur/be present than not." Its UD form can be stated as "If E occurs/is present, then F will always occur/be present." Once more we utilize *Ep* and *Fp* for a probabilistic relationship, and *Ed* and *Fd* for a UD one.

Let us suppose that we collect two different samples to test the two distinct hypotheses. In the case of the probabilistic relationship, let us assume that we found two cases where *Ep* and *Fp* are present, and three cases where *Ep* is present but *Fp* is absent. **Figure 5** illustrates these scenarios.

In the case of a UD hypothesis, let us suppose that we found four cases where *Ed* and *Fd* are present, and one case where *Ed* is present but *Fd* is absent.

Similar to the scenario disused before, it can be seen again that when researchers find evidence that "practically falsifies" a hypothesis stating that "if *Ep* occurs/is present, then *Fp* is more likely to occur/be present than not," for all pragmatic purposes, such a hypothesis should be discarded, as the evidence in **Figure 5** shows that in most cases, it does not hold. Therefore, researchers should be inclined to conclude that there is no evidence supporting the assertion that *Ep* and *Fp* are in some way associated. On the other hand, **Figure 6** shows that although the UD relationship "If *Ed* occurs/is present, then *Fd* will always occur/be present" has been logically and strictly falsified (i.e., one observation falsified this hypothesis), for all practical and theoretical purposes, researchers could still affirm that *Ed* is likely to be a causal agent for *Fd*, as the evidence shows that in most cases (four to one), the expected relationship does hold. Therefore, the falsified UD hypothesis between *Ed* and *Fd* can now be restated as a probabilistic hypothesis: "if *Ed* occurs/is present, then *Fd* is more likely to occur/be present than not."

The previous discussions show a clear difference between UD and probabilistic hypotheses: a "practically falsified" probabilistic hypothesis (*when Ho establishes a noticeable and positive relationship between the variables*) has no theoretical/empirical utility. On the other hand, a (strictly and logically) "falsified" UD hypothesis could still have pragmatic/empirical utility. In Section 5, we elaborate on the potential utility of hypotheses that have been falsified.

Response to criticism (b): a "practically falsified" probabilistic hypothesis (when Ho establishes a noticeable and positive relationship between the variables), for all theoretical/practical purposes, does not have predictive power/utility. On the other hand, a falsified UD hypothesis can still have predictive power/practical utility. Therefore, it is not a red herring to say that UD and probabilistic hypotheses are different in nature.

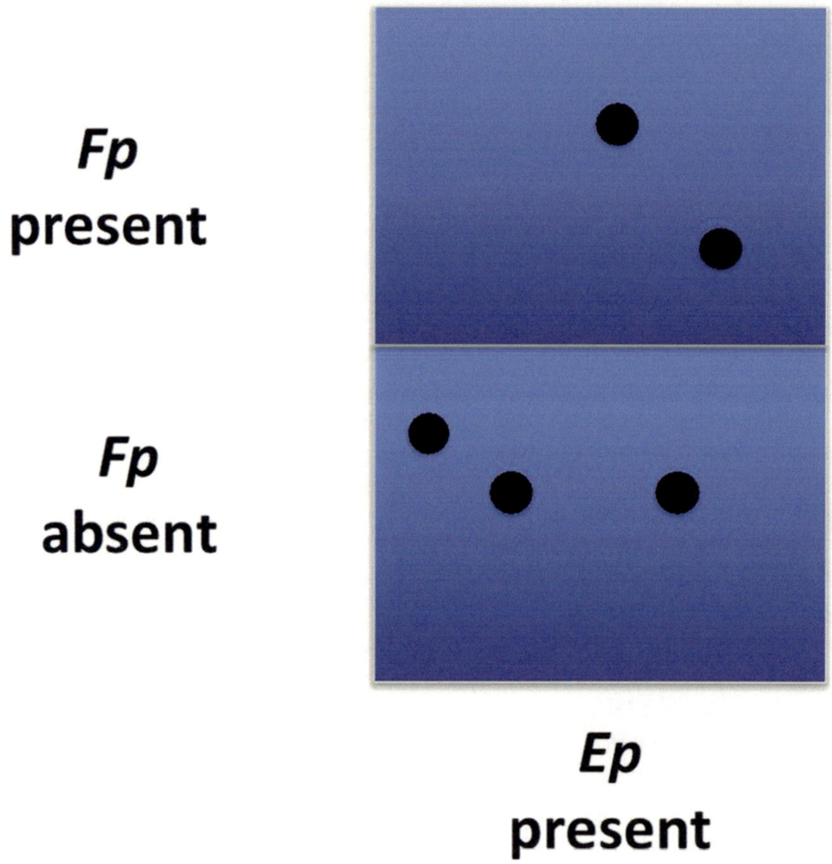

Figure 5.
Evidence that practically falsifies a probabilistic relationship between Ep and Fp.

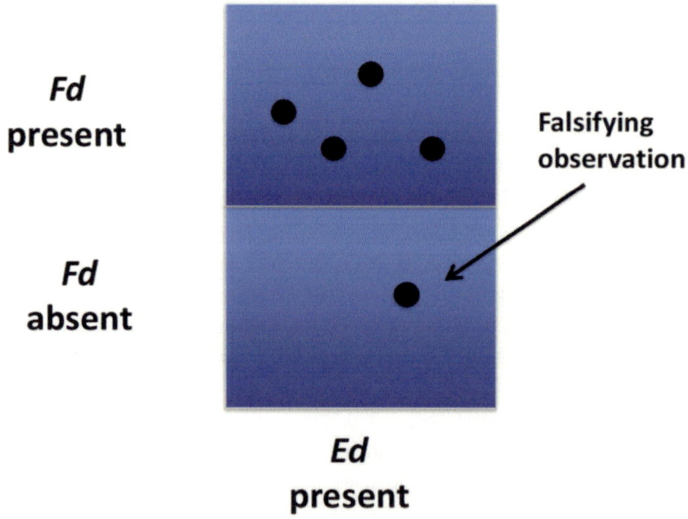

Figure 6.
Evidence that falsifies (logically and strictly) a UD relationship between Ed and Fd.

DOI: http://dx.doi.org/10.5772/intechopen.1000218

4. Criticism (c): "Few interesting relationships in operations management can be formulated as deterministic hypotheses"

The Cambridge Dictionary online[4] defines the word "interesting" as:

> Someone or something that is interesting keeps your attention because he, she, or it is unusual, exciting, or has a lot of ideas:

It could be validly argued that a UD relationship (e.g., something that always occurs) is far more interesting that something that takes place, for example, in most cases (e.g., a probabilistic relationship). Nonetheless, irrespective of why a researcher may deem a causal relationship interesting, we think that all scientists should be knowledgeable about the theoretical/empirical differences between UD and probabilistic hypotheses. This is important in order to avoid inaccuracies/omissions when hypothesis-testing and also when offering advice about research methodology (see [2] for a more detailed discussion about these issues in OM research).

Furthermore, while it is clear that the OM field (and arguably most of Social Sciences[5]) has been dominated by what could be called as "significance testing" [13], recently some scientists have also attempted to emphasize the importance and utility of UD hypotheses, for example, in the Business/Operations Management field [1, 14]. Likewise, Dienes ([8], p. 26, his italics) affirms that.

> Maybe many theories in psychology could effectively be written in the form, "In certain contexts, people always use this mechanism": "When my experimental procedure is set up in this way, all learning involves this sort of neural network."

Therefore, if UD hypotheses are to become more prominent in the Social Sciences, then it follows that scientists should be aware of their differences vis-à-vis probabilistic hypotheses, regardless of whether scientists (in their subjective views) think them interesting or not.

Response to criticism (c): whether a scientific proposition appears to be interesting or not, it is important for all scientists to be knowledgeable vis-à-vis the differences in the logical form and theoretical/empirical implications of UD and probabilistic hypotheses.

5. To be clear false theories can still be of use!

In Section 3, we explained that when a probabilistic hypothesis has been "practically falsified," researchers should be inclined to conclude that there is no evidence at all of a causal relationship between an independent and a dependent variable. We also explained that when a UD hypothesis has been logically and strictly falsified, it could still be of practical use, as a noticeable causal relationship between two variables could still be identified (see **Figures 4** and **6**).

It is not possible to be certain regarding the reasons that explain why expert researchers (see footnote 1) still opine that the differences between UD and probabilistic hypotheses are just a red herring and unimportant. Nevertheless, we think that a contributing factor might be the lack of awareness about the fact that falsified hypotheses can still be

---

[4] See https://dictionary.cambridge.org/dictionary/english/interesting [accessed 20 May 2022].

[5] See for example Hartmann and Sprenger [13] for a more detailed discussion on what they call "the mathematization of the social sciences."

of practical utility. It would appear as if researchers tend to equate the terms "falsified/false" with "useless," "meaningless" or "pointless." To support our contention that falsified hypotheses can still be of use, we quote Popper ([15], p. 74):

> ... false theories often serve well enough: most formulae used in engineering or navigation are known to be false, although they may be excellent approximations and easy to handle; and they are used with confidence by people who know them to be false.

Even though Newtonian physics have been "recognized as literally false" ([8], p. 5), "... scientists needed nothing more than Newton's equations to plot the course of the rocket that landed men on the moon" [16].

More specifically, in the OM field, the influential strategic trade-offs model [17] was put forward as a UD theory [2]. Even if empirical evidence were to falsify – logically and strictly – this proposition in the future, it would be difficult for researchers to affirm that Skinner's core argument ("*no product/service can be the best at everything*") does not reflect phenomena that occurs in the marketplace[6]. In short, even if Skinner's model were to be falsified, we argue that it would still be useful/have practical utility in the understanding – at least in some circumstances/situations – of the differences observed in the features/performance between pairs of rival products/services.

The fact that strictly and logically false/falsified (UD) hypotheses can still be of practical/empirical use should serve to strengthen our argument about the importance of acknowledging the differences that exist between UD and probabilistic hypotheses in empirical science.

6. Some concluding remarks

Investigations that deal directly with the topics of universal-deterministic and probabilistic hypotheses are not common in business management research in general, and operations management science in particular. Our hope is that this paper can contribute to a better understanding of the theoretical/empirical differences that exist between these two types of hypotheses.

[6] See [18] for a detailed discussion of strategic trade-offs between two competing products. Moreover, some authors (e.g., [19]) claim that in practice, the trade-offs model is not used. Sarmiento et al. [20] provide a detailed response to Singh et al.'s assertion.

Author details

Roberto Sarmiento
Faculty of Engineering, Autonomous University of San Luis Potosi, Mexico

*Address all correspondence to: sarmientoroberto@yahoo.com.mx

References

[1] Dul J, Hak T. Case Study Methodology in Business Research. Oxford UK, and Burlington, MA, USA: Butterworth-Heinemann; 2008

[2] Whelan G, Sarmiento R, Sprenger J. Universal-deterministic and probabilistic hypotheses in operations management research: A discussion paper. Production Planning & Control. 2018;**29**(16):1306-1320

[3] Ross D. I don't Understand Quantum Physics. Available from: https://www.southampton.ac.uk/~doug/quantum_physics/quantum_physics.pdf. Southampton, UK: University of Southampton; 2018 [Accessed: May 20, 2022]

[4] Eisenhardt KM. Building theories from case study research. Academy of Management Review. 1989;**14**(4):532-550

[5] Voss C, Tsikriktsis N, Frohlich M, M. Case research in operations management. International Journal of Operations and Production Management. 2002;**22**(2):195-219

[6] Yin RK. Case Study Research: Design and Methods. London: Sage; 2014

[7] Popper K. The Logic of Scientific Discovery. London and New York: Routledge Classics, Taylor & Francis Group; 2002

[8] Dienes Z. Understanding Psychology as a Science: An Introduction to Scientific and Statistical Inference. Basingstoke and New York: Palgrave-Macmillan; 2008

[9] Folsing A. Albert Einstein: a biography. Viking (published by the Penguin Group), London: Penguin Books Ltd; 1997

[10] Overbye, D. 2012. The Trouble With Data That Outpaces a Theory. Available from: http://www.nytimes.com/2012/03/27/science/the-trouble-with-neutrinos-that-outpaced-einsteins-theory.html. [Accessed: December 13, 2015]. Available from: https://www.nytimes.com/2012/03/27/science/the-trouble-with-neutrinos-that-outpaced-einsteins-theory.html. [Accessed: May 18, 2022]

[11] Cartlidge E. BREAKING NEWS: Error Undoes Faster-Than-Light Neutrino Results. 2012. Available from: http://news.sciencemag.org/2012/02/breaking-news-error-undoes-faster-light-neutrino-results?ref=hp. [Accessed: December 13, 2015]. Available from: https://www.science.org/content/article/breaking-news-error-undoes-faster-light-neutrino-results. [Accessed: May 18, 2022]

[12] Miller D. Popper's contributions to the theory of probability and its interpretation. In: Shearmur J, Stokes G, editors. The Cambridge Companion to Popper. Cambridge: Cambridge University Press; 2016. pp. 208-229

[13] Hartmann S, Sprenger J. Mathematics and statistics in the social sciences. In: Jarvie IC, Zamora Bonilla J, editors. The SAGE Handbook of the Philosophy of Social Sciences. London: SAGE; 2011. pp. 594-612

[14] Dul J, Hak T, Goertz G, Voss C. Necessary condition hypotheses in operations management. International Journal of Operations & Production Management. 2010;**30**(11):1170-1190

[15] Popper K. Conjectures and Refutations: The Growth of Scientific Knowledge. London and New York:

Routledge Classics, Taylor & Francis Group; 2002

[16] Green B. The Elegant Universe: Episode 1 - Einstein's Dream. Available from: https://www.youtube.com/watch?v=5vUzzCoArCo. [Accessed: May 20, 2022]

[17] Skinner W. Manufacturing-missing link in corporate strategy. Harvard Business Review. 1969;**47**(3):136-145

[18] Sarmiento R, Thurer M, Whelan G. Rethinking Skinner's model: Strategic trade-offs in products and services. Management Research Review. 2016;**39**(10):1199-1213

[19] Singh PJ, Wiengarten F, Nand AA, Betts T. Beyond the trade-off and cumulative capabilities models: Alternative models of operations strategy. International Journal of Production Research. 2015;**53**(13): 4001-4020

[20] Sarmiento R, Whelan G, Thürer M. A note on 'beyond the trade-off and cumulative capabilities models: Alternative models of operations strategy'. International Journal of Production Research. 2018;**56**(12): 4368-4375

Chapter 2

Design and Planning Robust and Competitive Supply Chains

Kotomichi Matsuno, Jiahua Weng, Noriyuki Hosokawa and Takahiro Ohno

Abstract

In recent years, supply chains in the manufacturing industry have become more and more complicated, and many cases of supply chain disruptions due to natural disasters have been confirmed. It is necessary for manufacturers to build a system that can help them alleviate losses and shorten recovery periods due to supply chain disruptions. Supplier diversification, as well as supplier evaluation and selection, are discussed as risk aversion measures in many papers. However, even if the procurement source has been evaluated enough, there are problems, such as opportunity loss during recovery periods and soaring procurement costs during normal periods. In this chapter, to help Japanese manufacturers to alleviate opportunity loss under component procurement disruption situations and keep cost competitiveness in normal periods, decision-making models of supply chain structure assessment, supplier selection, procurement allocation, and trading contracts are designed and verified.

Keywords: supply chain disruption, supply chain structure, supplier selection, procurement allocation, contract

1. Introduction

The supply chain structure in the manufacturing industry is getting more and more complex due to the acceleration of globalization. It makes the impact of disruptions on the whole supply chain greater than ever before. Many natural and man-made disasters, such as Hurricane Katrina 2005, Great East Japan Earthquake 2011, Thailand Flood 2011, Kumamoto Earthquake 2016, COVID-19 2019, and the recent Russia-Ukraine war 2022, have resulted in large-scale supply chain disruptions. Manufacturers had a large amount of opportunity losses due to production interruptions caused by the shortage of components. Moreover, a number of manufacturers were unable to get survived anymore due to component procurement disruptions.

In the past, how to reduce procurement costs, shorten procurement lead time, and maintain good relationships with suppliers are considered the most important matters when manufacturers make their decisions on procurement. Since the Great East Japan Earthquake of 2011, the development of an effective business continuity plan (BCP) has been considered one of the most important matters to Japanese manufacturers. As a part of BCP development, sourcing decision-making with consideration of supply chain disruption risks is becoming the most important process. In the increasingly competitive global environment, manufacturers should not only focus on the efficiency of their own production and logistics operations but also take measures, such as understanding the location of risks in the supply chain and diversifying procurement sources to reduce the risks of procurement shortages due to supply chain disruptions [1].

As known to all, BCP defines the activities for business continuity. It should be carried out under both normal and emergency situations to enable the business continuation or early recovery of core business operations in the event of natural disasters, major fires, terrorist attacks, or other emergency situations. With the aim of increasing the development rate of BCP in Japan, the Small and Medium Enterprise Agency has published a BCP development manual on its website, which explains how to develop a BCP for each business in detail. The process of BCP development is shown in **Figure 1** [2].

There are five steps for BDP development and operation, including (1) understand business operations, (2) consider BCP preparations and prior measures, (3) formulate BCP, (4) establish BCP culture, and (5) test, maintain, and update BCP. Step 2 and step 3 are considered the most important and difficult steps among all the five steps. The significance of BCP and its economic effects are mentioned to show that BCP should be a system that also could generate economic effects even in normal periods compared to conventional disaster prevention measures [3]. However, many manufacturers are stuck between step 2 and step 3. Lacking visibility of benefits under normal situations

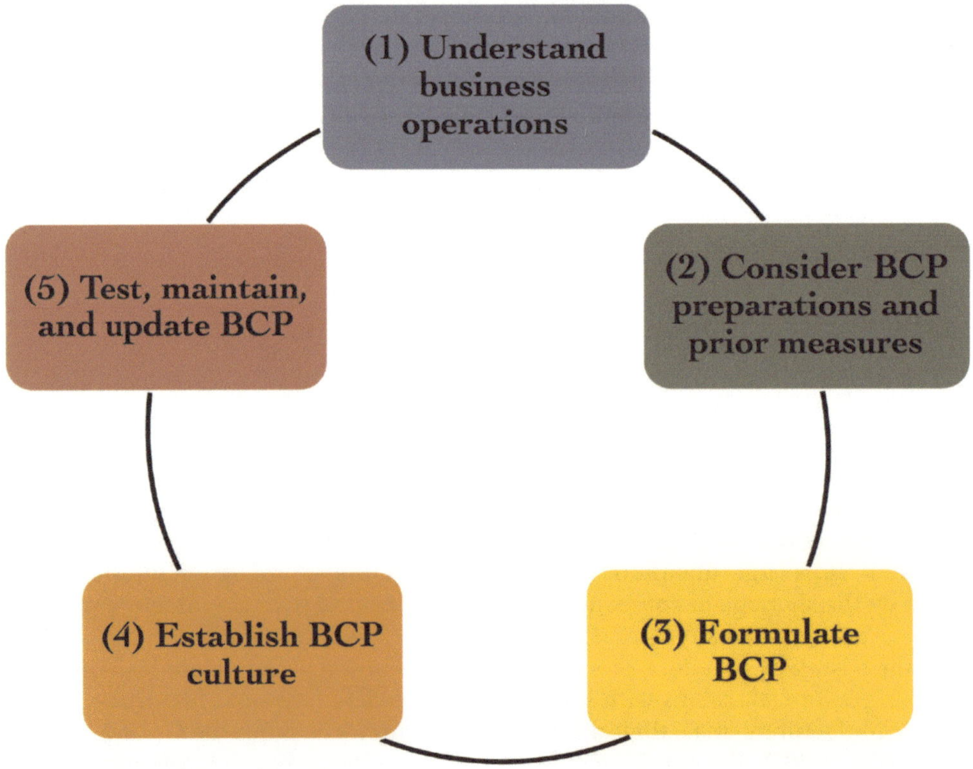

Figure 1.
BCP development and operation cycle.

and lacking mitigation effects against losses due to supply chain disruptions are cited as the main reasons for lacking progress in BCP development.

To help manufacturers to make their decision-making on risk mitigation measures creation as well as effectiveness evaluation, research on supply chain disruption risk management (SCDRM) to deal with supply chain disruptions has been attracting much attention and the number of papers has been increasing rapidly in recent 10 years. Proposals of these papers could be broadly divided into two types which are pre-measures and post-measures. Risk assessment of supply chain structures, supplier evaluation, and procurement allocation is considered the pre-measures. On the other hand, switching to alternative components and switching procurement sources are considered the post-measures. In addition, since most of the post-measures are ineffective without any pre-measures, pre-measures can be considered as plans for post-measures.

Based on the levels of decision-making, risk mitigation measures could be divided into strategic level, tactical level, and operational level. Assessing supply chain risks and determining what kind of supply chain structure should be built is considered the strategic level of decision-making. As a part of supply chain building, evaluating and selecting suppliers based on the decided supply chain structure is also considered the strategic level of decision-making. Concluding risk-hedging trading contracts, including procurement agreements, supply and penalty terms between upstream and downstream companies are considered the tactical level of decision-making. How to make post-measures more efficient is considered an operational level of decision-making. **Figure 2** shows the risk mitigation measures by different decision-making levels. In this chapter, decision-making models of risk mitigation measures at the strategic level and tactical level are mainly discussed.

The remainder of this chapter is organized as follows. Section 2 provides an introduction to the risk assessment of supply chain structures. Section 3 provides a study on the decision-making of manufacturers on suppliers selection and procurement allocation based on the risk assessment of supply chain structures. Section 4 presents a robust and competitive contract model for manufacturers. The chapter is concluded in Section 5.

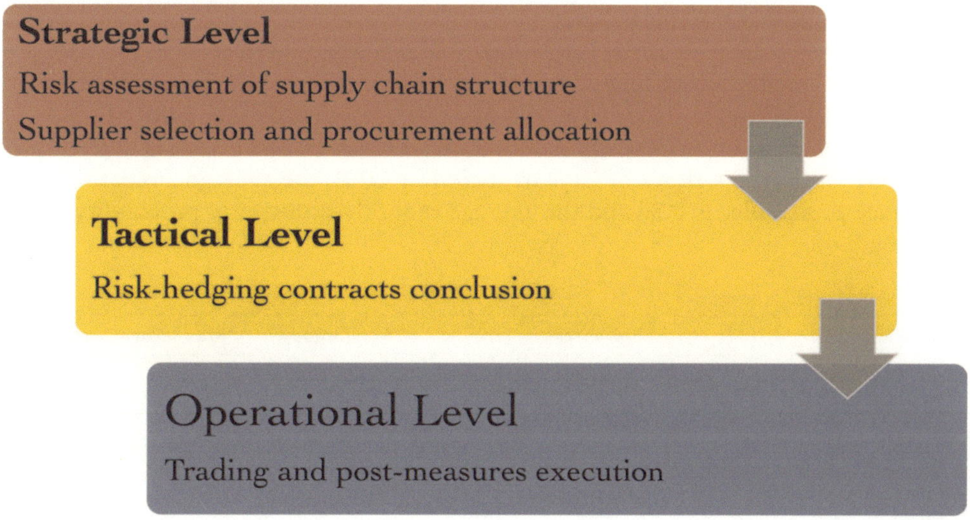

Figure 2.
Risk mitigation measures at different decision-making levels.

2. Risk assessment of supply chain structure

This section presents literature reviews and a previous study about risk assessment of supply chain structures [4].

The economic impact of supply chain disruptions immediately after the Great East Japan Earthquake was examined and the amount of production losses caused by the supply chain disruptions to at least 0.35% of the GDP was estimated in [5]. In ref. [6], the responses of Japanese manufacturing firms to natural disasters, such as earthquake, tsunami, and nuclear disasters, were discussed. Based on case studies, several supply-chain recovery processes after natural disasters as well as humanitarian disruptions were discussed and reflection points in terms of disaster planning, and recovery responses were summarized. In ref. [7], several supply-chain structural models were built and the tradeoff relation between supply-chain efficiency and robustness under supply-chain disruption risks was discussed. Countermeasures for mitigating supply-chain disruption risks in terms of redundancy, robustness, and flexibility were discussed after the research work on supply-chain visualization in ref. [8].

As typical natural disasters, earthquakes cause significant damage to supply chains. Since earthquakes occur frequently in Japan, procurement disruptions due to earthquakes also occur frequently. Occurrence probability, as well as impact, are two main evaluation indicators of earthquakes. In this study, the recovery period of the supply chain from the earthquake is considered as the earthquake impact. A model of procurement disruptions due to earthquakes is developed to clarify the loss difference among different supply-chain structures in Japan.

2.1 Model of procurement disruptions due to natural disasters

Although the probability of an earthquake cannot be expressed accurately, as a general thought, occurrence probability could be estimated based on historical data on earthquakes. Since earthquakes less than seismic intensity 7 only have little impact on production and procurement activities, only earthquakes with a seismic intensity of 7 or over are considered in this study.

Table 1 shows the data about earthquakes over seismic intensity 7 in the last 10 years (2011–2021) which was collected from the website of the Japan Weather Association [9]. With the data in **Table 1**, the average annual probability of an earthquake which seismic intensity is 7 could be estimated. Since seismic intensity 7 earthquakes occurred 4 times in 11 years including 2021, the average annual occurrence probability is 0.36 and the average monthly occurrence probability should be 0.03.

Date and time	Epicenter	Magnitude	Max seismic intensity
2018/9/6 3:08	Eastern Iburi of Hokkaido	M6.7	7
2016/4/16 1:25	Kumamoto region of Kumamoto	M7.3	7
2016/4/14 21:26	Kumamoto region of Kumamoto	M6.5	7
2011/3/11 14:46	Sanriku offshore	M7.9	7

Table 1.
Earthquakes over seismic intensity 7 in the last 10 years (2011–2021) in Japan.

In addition, the number of related areas is an important attribute of the supply chain when discussing the probability of supply-chain disruptions. In general, the larger the number of related areas of a supply chain, the higher the probability of disruption will be. Japan is divided into seven regions which are used for statistical by government officials. They are Hokkaido, Tohoku, Kanto, Chubu, Kinki, Chugoku-Shikoku, and Kyushu-Okinawa. In this study, the relation between a number of related areas of component procurement activities and the probability of disruptions is focused to verify the disruption impact. **Table 2** shows the disruption probability according to the number of related areas of a supply chain. It is calculated with a number of regions and the average monthly occurrence probability of earthquakes over seismic intensity 7.

The recovery period of the supply chain from the earthquake could also be estimated based on past earthquake occurrences. Generally, the recovery period from earthquakes depends on industry types; it is necessary to estimate the recovery period with the data which should be collected from the manufacturing industry. **Table 3** shows the recovery periods of several electronic device manufacturers after the Kumamoto earthquake on April 16, 2016 [10].

The date of partial recovery in the table is the date when production activities are restarted partially. The date of full recovery in the table is the date when production activities recovered as the same as the production capacity planned beforehand. It is difficult to show the recovery of production capacity over time without the information on the recovery speed. Furthermore, it is also difficult to collect information on the recovery speed of production capacity since it is considered confidential information to manufacturers. To simplify this problem, in this study, only two conditions which are "recovered" and "not recovered" are taken into consideration, and then the average period of partial and full recovery of all manufacturers according to **Table 3** is set as the estimated recovery period. With the results of the calculation, the estimated recovery period should be 36.6 days, nearly 1 month after the earthquake.

Related areas	**1**	**2**	**3**	**4**	**5**	**6**	**7**
Disruption probability	0.004	0.009	0.013	0.017	0.022	0.026	0.030

Table 2.
Disruption probability by related areas of supply chain.

Business office	**Location**	**Date of partial recovery**	**Date of full recovery**
Renesas Semicond. Manuf. Co., Ltd., Kawashiri factory	Kumamoto-shi	Apr. 22	May 22
FUJIFILM Kyusyu Co.,Ltd.	Kikuyomachi	Apr. 23	May 23
Mitsubishi Electric Corp. Power Device Works Kumamoto	Koushi-shi	May 9	May 31
Tokyo Electron Kyushu Ltd.	Koushi-shi	Apr. 25	End of June
Sony Semicond. Manuf. Co., Ltd., Kumamoto Tech. Ctr.	Kikuyomachi	May 9	End of July

Table 3.
Recovery dates of electronic device manufacturers after Kumamoto earthquake 2016.

2.2 Supply chain structure models

How to build and maintain competitive supply chains is the most important task for manufacturers. In particular, manufacturers whose products are in the competition period, such as smartphones in Japan market in the 2010s, LCD TV and PC in Japan market in 2000s.

In general, manufacturers adjust their suppliers and procurement terms for the next year annually based on their strategies and evaluation results of performances in the last year. In the past, procurement plans were decided by manufacturers based on an overall evaluation of procurement cost, quality of the component, and delivery time. In most cases, the minimum number of suppliers at the lowest procurement costs was decided as the procurement plan for next year. **Figure 3** shows a basic supply chain structure which is a one-supplier-one-manufacturer structure. It is considered an extreme model of a few sources supply chain. In an electronic products supply chain, such as LCD TV and smartphone, comparing with the order receiving cycle of finished products, the procurement lead time of electronic components such as LCD panel is much longer. With reference to many LCD TV and smartphone assembly manufacturers, in this supply-chain model, the order receiving cycle of products from retailers is set to 1 month, lead time of component procurement from the supplier is set to 3 months.

To analyze the impact of components procurement disruptions on the supply-chain structure, profit of the manufacturer should be the most important indicator. Eq. (1) shows the profit function of manufacturer. It is the difference between sales amount and costs including component procurement cost, inventory cost, and production cost.

$$\pi_m = \sum_{n=1}^{N} \Big\{ p_p \cdot \min\left(q_{p.n}, I_{n-1}\right) - \beta_n c_{proc.}\, q_{c.n} - c_{inv.}\left(I_{n-1} + \beta_n q_{c.n} - q_{p.n}\right) - \left[c_f + c_v \cdot \min\left(q_{p.n}, I_{n-1}\right)\right] \Big\} \quad (1)$$

p_p: Product price.

$q_{p.n}$: Volume of product orders from the retailer.

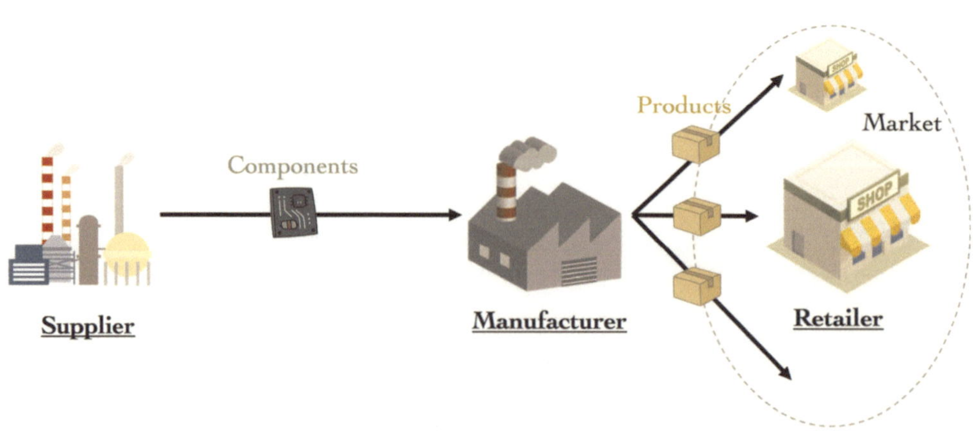

Figure 3.
A simple centralized procurement structure.

I_n: Inventory level of components.
$c_{proc.}$: Component procurement cost/per unit.
$q_{c.n}$: Volume of component procurement.
$c_{inv.}$: Inventory cost/per unit.
c_f: Fixed cost of production.
c_v: Variable cost of production.
N: Period (Month).
β_n: If there is no procurement disruption, $\beta_n = 1$. Otherwise $\beta_n = 0$.

Such centralized procurement structures can help manufacturers to reduce procurement costs in normal periods. However, manufacturers are not able to achieve components in the recovery periods when a supply chain disruption occurs. It makes manufacturers a huge opportunity loss due to production stoppages. As a general robust countermeasure for hedging disruption risks, multiple sourcing has been used widely. **Figure 4** shows a simplified robust procurement structure with one more supplier adding to **Figure 3**. If one of the procurement channels gets disrupted, the manufacturer can receive part of the total components from another supplier at least. The with this procurement structure. The manufacturer is able to keep its production facility utilization to a certain degree and avoid some opportunity losses. The profit of the manufacturer in this structure is shown in Eq. (2).

$$\begin{aligned}\pi_m = & \sum_{n=1}^{N}\Big\{p_p \cdot \min\left(q_{p.n}, I_{n-1}\right) - \gamma_n c_{proc.}\left(q_{c1.n} + q_{c2.n}\right) \\ & - c_{inv.}\left[I_{n-1} - q_{p.n} + \gamma_n\left(q_{c1.n} + q_{c2.n}\right)\right] - \left[c_f + c_v \cdot \min\left(q_{p.n}, I_{n-1}\right)\right]\Big\}\end{aligned} \tag{2}$$

$q_{p.n}$: Volume of product orders from the retailer
$q_{c1.n}$: Volume of component procurement from supplier 1
$q_{c2.n}$: Volume of component procurement from supplier 2
γ_n: If there is no procurement disruption, $\gamma_n = 1$. Otherwise $\gamma_n = k$
k: If component procurement from supplier 2 is disrupted $k = q_{c1.n}/q_{p.n}$, If component procurement from supplier 1 is disrupted $k = q_{c2.n}/q_{p.n}$. If component

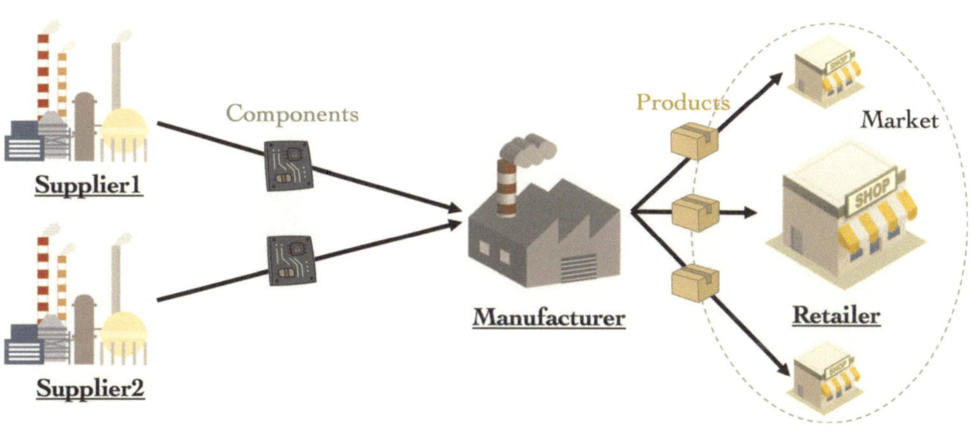

Figure 4.
A simple decentralized procurement structure.

procurement from both suppliers is disrupted $k = 0$, since the probability of such a case is extremely low, it is not considered in the experiment of this study.

In general, costs such as supplier management fees, contract fees, and ordering costs increase with the number of suppliers increasing. Although this decentralized procurement structure can be expected to reduce opportunity losses in the event of supply chain disruptions, the cost competitiveness of the supply chain is also decreased due to higher operating costs during normal periods.

2.3 Numerical experiment and results

Attributes of the supply chain, including related areas of procurement, are discussed to analyze the profit of manufacturer to clarify the difference between centralized procurement structure and decentralized procurement structure. The period of experiments is set to 3 years, and data for experiments is shown below.

Estimated product demand/month: 1000
Component procurement cost: 10,000
Component order lot size: 3000
Production cost (fixed): 10,000,000
Production cost (variable): 1000
Component procurement lead time: 3 months
Actual product demand/month: $N(1000, 200^2)$
Product price: 35,000
Inventory cost/month: 200

Figure 5 shows the total profit of the manufacturer which is the average value of 50 simulation runs with the number of related areas from 1 to 5. The vertical axis of

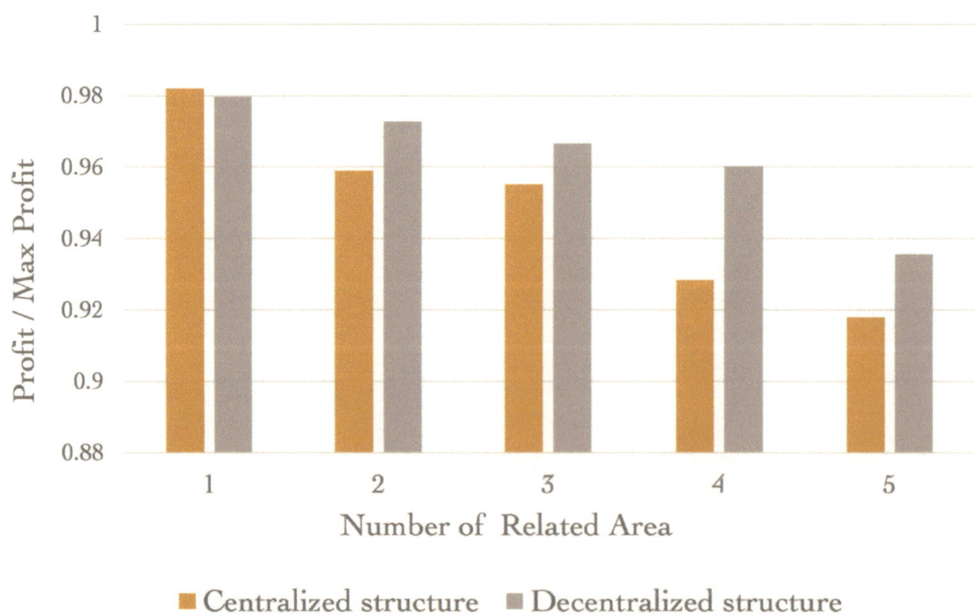

Figure 5.
Total profit of manufacturer by different related areas.

the graph is the ratio of profit to maximum profit which could be achieved in a centralized procurement structure without any procurement disruptions in 3 years. From **Figure 5**, we can find that in more related areas of component procurement, lower profits for manufacturers in both centralized and decentralized procurement structures would be. However, comparing with a centralized structure, no significant profit reduction with increasing in related areas with a decentralized procurement structure. The effect of opportunity loss avoidance of a decentralized procurement structure is confirmed. The peak profit difference between the two structures could be confirmed in **Figure 5** with the trend of increase and decrease.

3. Supplier selection and procurement allocation

Supplier selection and procurement allocation are considered one of the most important strategy level decisions to manufacturers in planning component procurement. Generally, the decision-making of supplier evaluation and selection is based on the determined supply-chain structure. This section presents literature reviews and previous work on supplier selection and procurement allocation [11].

The procurement evaluation problem which was dealt with as a multi-objective function optimization problem with randomly varying weights of cost, quality, delivery, as well as economic risks was discussed in ref. [12]. Supplier selection from all candidate suppliers by comprehensive scale assessment, including procurement risks and costs is shown in ref. [13]. In ref. [14], the problem of an optimal number of suppliers selection by considering the component inventory level under procurement lead-time uncertainty is discussed. In [15], a supplier selection method for the purpose of minimizing both procurement costs and opportunity losses was proposed. The method is based on the assumption that the component procurement price of each supplier is the same, and the order volume from each supplier is also the same. Reference [16] focused on the decoupling of recurrent supply risks and disruption risks during sourcing decisions and procurement allocation.

As decision-making of supplier selection and procurement allocation based on the decentralized procurement structure, minimum total procurement costs, including management cost, purchasing cost, transportation cost, and the opportunity cost with considering procurement disruption should be set as the objective function.

3.1 Objective functions and constraints

An optimization method for supplier selection and procurement allocation based on a one-manufacturer-multiple-suppliers supply-chain structure is proposed in this study. It is the decision-making of the manufacturer on where and how many components to procure from candidate component suppliers whose basic attributes, such as quality and delivery could meet the requirements of the manufacturer for the next period. Therefore, decision variables are set as below.

n : Number of suppliers
ZX_s: If components are sourced from the supplier s, $ZX_s = 1$. Otherwise $ZX_s = 0$.
Q_{is} : Procurement volumes of components i from supplier s

Objective function

$$PC_{min} = Mgmt.Cost + Pur.Cost + Trans.Cost + Oppty.Loss \quad (3)$$

As discussed before, minimum total procurement costs, including management cost, purchasing cost, transportation cost, and the opportunity cost with considering procurement disruption, are considered as the objective function. Management cost is the cost involved in managing supplier activities and contracts. It increases or decreases depending on the number of suppliers. The purchasing cost is the total cost of purchasing components from all suppliers. Transportation costs can be calculated by multiplying the unit transportation cost of each component from one supplier to the manufacturer by transportation quantity which is the same as purchasing quantity.

$$Fixed\ Management\ Cost = \sum_{s=1}^{S}\sum_{i=1}^{I} ZX_s \times M_s \tag{4}$$

$$Purchasing\ Cost = \sum_{s=1}^{S}\sum_{i=1}^{I} P_{is} \times Q_{is} \tag{5}$$

$$Transportation\ Cost = \sum_{s=1}^{S}\sum_{i=1}^{I} Trans_{is} \times Q_{is} \tag{6}$$

Eq. (7) shows the amount of opportunity loss due to any n procurement channels getting disrupted. Eq. (8) shows the expected total opportunity cost which is the sum of Eq. (7)

$$C_n = \sum_{i=1}^{I} L_i \times \sum \left\{ \left[D_i - \min\left(D_i, \sum_{b=1}^{B} Q_{ib}^{max} \right) \right] \times \left[\prod_{a \in A_n} \theta_a \times \prod_{b \in B_n} (1 - \theta_b) \right] \right\} \tag{7}$$

$$\text{Oppotunity Loss} = \sum_{n=1}^{S} C_n \tag{8}$$

Constraints

Equations from Eq. (9) to Eq. (11) are considered as the constraints for decision-making, including total demand for components, procurement volume from each supplier, and the number of suppliers. Eq. (9) shows the constraint of component demand. The total volume of component i that could be procured from all selected suppliers should be the same as the demand for component i. Eq. (10) shows the constraint of component volume. The volume of component i that could be procured from supplier s should not be less than the minimum volume and not more than the maximum volume which is decided by supplier s. Eq. (11) shows the constraint of the number of suppliers available for selection. It should not exceed the number of all candidate suppliers.

$$\sum_{s=1}^{S} Q_{is} = D_i \forall i \tag{9}$$

$$Q_{is}^{min} \leq Q_{is} \leq Q_{is}^{max} \quad \forall i, s \tag{10}$$

$$n \leq S \tag{11}$$

s : Supplier ID ($s = 1, 2, 3, \ldots, S$)
i : ID of component types ($i = 1, 2, 3, \ldots, I$)

D_i : Demand for component i
M_s : Management cost of supplier s
P_{is} : Price of component i from supplier s (per unit)
Q_{is}^{min} : Minimum procurement volume of components i from supplier s
Q_{is}^{max} : Maximum procurement volume of components i from supplier s
L_i : Opportunity loss of components i (per unit)
$Trans_{is}$: Transportation cost of components i from supplier s (per unit)
θ_s : Disruption probability of supplier s (%)
a : ID of interrupted suppliers among selected suppliers ($a = 1, 2, 3, \dots, A$)
b : ID of among selected suppliers ($b = 1, 2, 3, \dots, B$)
A_n : The interrupted supplier group (Ex. A_1 indicates the group of anyone interrupted supplier)
B_n : The uninterrupted supplier group (Ex. B_1 indicates the group of uninterrupted suppliers when any supplier is interrupted)
Q_{ib}^{max} : Maximum procurement volume of components i from uninterrupted suppliers

3.2 Tabu-search based decision-making

In many cases, procurement plans are decided based on the discussion in a procurement strategy meeting. As a support tool for management decisions, create and re-create supplier selection and procurement allocation plans in a short time are important. The effectiveness of Tabu-Search (TS) has been confirmed by numerous previous studies on a lot of classical optimization problems as well as practical problems. Comparing with other meta-heuristics, it is expected to search more neighborhood solutions with fewer search times. Since prohibitions are introduced to discourage search from coming back to previous-visited solutions. With considering of the applicability, a TS-based decision-making is proposed. There are five steps to decide the optimal supplier combination and procurement volume from each supplier. In this study, the average procurement cost of the supplier combination is considered as the evaluation index.

Step 1: Generate and check the initial supplier combination

Initial supplier combination according to the tentatively determined numbers of suppliers is generated randomly. The procurement amount of each component from each supplier in the initial supplier combination is set randomly based on the constraint Eq. (10).

Step 2: Check and generate neighbor solutions

Check whether or not the solution could be improved according to the evaluation index. If it could be improved, create new supplier combinations by selecting the unselected suppliers in all candidate suppliers to replace the supplier with the highest procurement cost in the supplier combination after checking the constraints as well as the Tabu list. Otherwise, proceed to Step 4.

Step 3: Update the Tabu list

Update the Tabu list by adding the supplier combination with the lowest unit cost from supplier combinations generated by Step 2.

Step 4: Adjust procurement allocation

Adjust the component volume from the supplier with the highest procurement cost to the supplier with the lowest component procurement cost by minimum adjustment unit.

Step 5: Update execution times

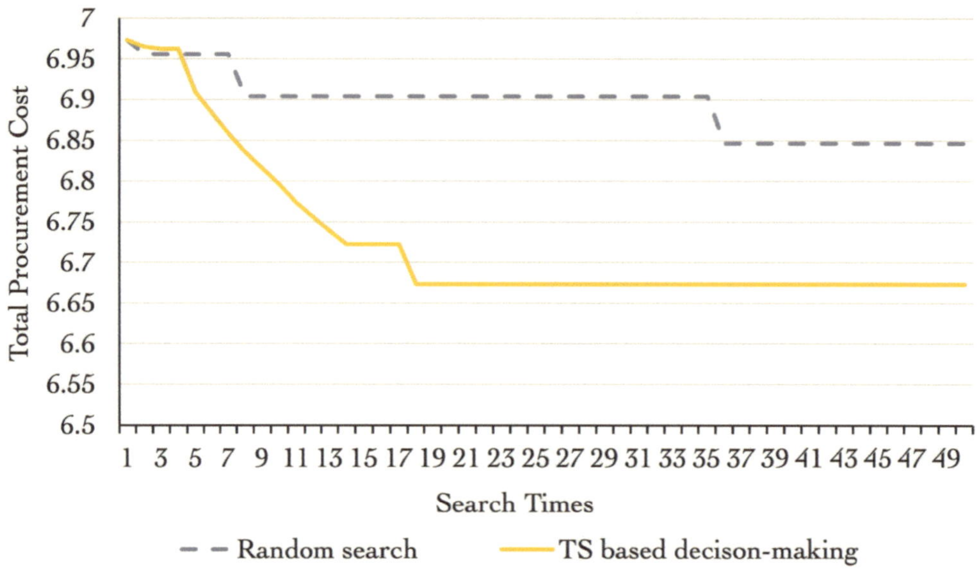

Figure 6.
Total procurement cost search with random search and TS-based decision-making based on three suppliers.

If the execution time *n* is less than the total search time *N*, proceed to Step 2. Otherwise, end the execution.

3.3 Numerical experiment and results

The effectiveness of the proposed TS-based decision-making is discussed in this subsection. The data for the numerical experiment is shown below.

Number of candidate suppliers: 8

Type of components: 3

Demand for each component type: 400, 300, 200

Range of management cost: 1500–2500

Minimum procurement volume range of each component type: 60–100, 45–75, 30–50

Maximum procurement volume range of each component type: 100–180, 80–150, 60–100

Component price range of each component type/per unit: 35–52, 40–62, 49–72

Transportation cost range of each component type/per unit: 3–7, 3–7, 3–7

Disruption probability of candidate suppliers: 18–27%

Figure 6 shows an example of the performance of TS-based decision-making against a random search in the terms of three suppliers. Comparing with random search, TS-based decision-making could reach a more accurate solution with much less search times. It could support managers to discuss and make a procurement plan efficiently in a management strategy meeting.

4. Design robust and competitive contract under disruption risks

With the results of risk assessment of supply-chain structures, as well as the results of supplier selection and procurement allocation, a trading contract between manufacturer and suppliers in which component procurement details are determined, should be discussed. In general, the procurement allocation determines the reference procurement volume of each component from each supplier. A trading contract that is concluded between manufacturer and supplier determines the terms of component trading. The manufacturer procures components from the supplier based on the estimated demand for a decided period. According to the example of LCD TV and smartphone assembly industry mentioned in Section 2, if the order-receiving cycle of products from retailers is 1 month and the lead time of component procurement from the supplier is 3 months, the manufacturer should place an order to the supplier with considering of the estimated demand for one month after the next 3 months and its component inventory.

Risk assessment of supply-chain structures and supplier selection and procurement allocation with risk evaluation can help manufacturers to reduce their procurement risks, but they could not help manufacturers to alleviate their losses in the event of a supply disruption. A robust and competitive procurement contract with post-measures execution consideration is also necessary for manufacturers. Many previous studies discussed the effectiveness of options contracts in managing supply-chain disruption risks. In ref. [17], a model for a three-echelon supply chain including two different manufacturers, one distributer and one retailer, to analyze the effect of the combination of option and buyback contracts under supply-disruption risk was created. In ref. [18], a supply-chain problem where a buyer receives a product from a cheap but unreliable main supplier and signs an option contract with a perfectly reliable backup supplier to share supply and demand uncertainty was discussed. A mechanism was developed to maximize system efficiency and a win–win relationship between supplier and buyer based on an optimization problem with option contracts. Reference [19] focused on a supply chain including a manufacturer who faces stochastic demand, an overseas module supplier, and two local module suppliers. The overseas supplier offers high-quality products while being susceptible to disruption risks. Though local suppliers have high reliability, it may decrease sharply if they offer products of lower quality. Optimal order volume and reservation volume of the manufacturer based on capacity reservation contract were derived.

The purpose of the contracts discussed above is to give procurement volume an adjustable range to avoid loss caused by uncertainties including supply disruptions. If this adjustable range is large enough, manufacturers can procure the necessary quantity of parts based on fluctuations in demand, and part of the loss caused by disruption can be covered by procuring the maximum level of the adjustable range from other suppliers.

4.1 Robust and competitive contract model

To realize such an adjustable range for procurement, it is necessary for manufacturers to make a lump-sum payment to suppliers to determine the upper limit of procurement volume in advance. On the other hand, the higher the upper limit of the procurement volume range is, the supplier has to bear more inventory risk. Generally, the lump-sum amount is decided by the supplier with considering the inventory risk based on the upper limit of procurement volume.

Figure 7 shows the structure of trading volume and price with the upper limit of procurement volume that is dependent on the amount of lump-sum payment. $Q_{ref.}$ is the reference volume decided by the manufacturer due to the decision on procurement allocation. Q_{max} is an upper limit of procurement volume. $PC(q)$ is the total

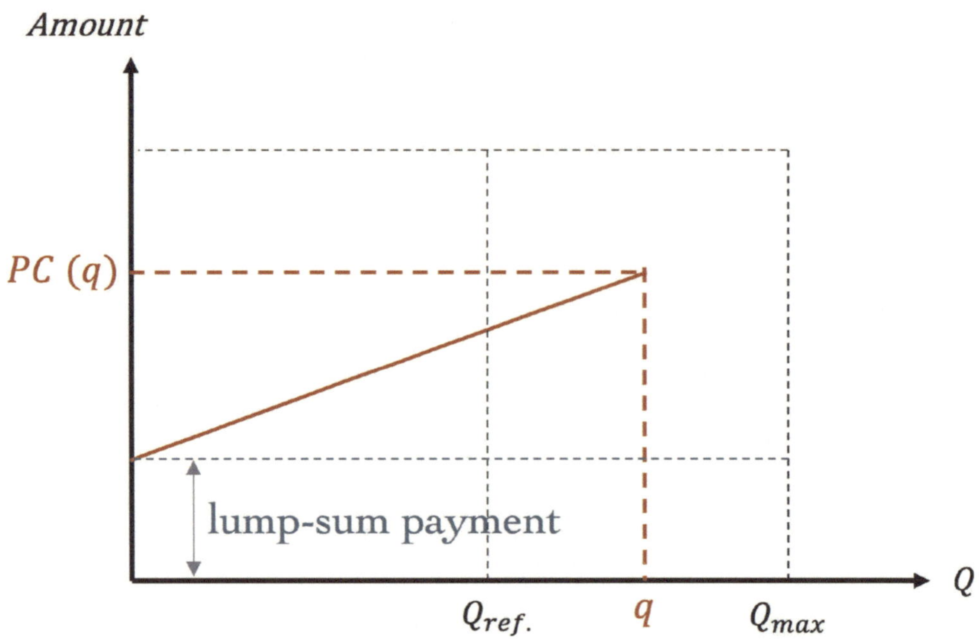

Figure 7.
Structure of trading volume and price with the lump-sum payment.

value of the lump-sum payment amount and $c_{proc.}$ which is the procurement amount with procurement volume q.

Eq. (12) shows the profit of the manufacturer with this contract structure based on the simple decentralized procurement structure, as shown in **Figure 4**. It is similar to Eq. (2). With the lump-sum payment ls_1 and ls_2 to supplier 1 and supplier 2, the manufacturer can make the decision on procurement volume with a wider adjustable range each time.

$$\begin{aligned}\pi_m = & \sum_{n=1}^{N}\Big\{p_p \cdot \min\left(q_{p.n}, I_{n-1}\right) - (ls_1 + ls_2) - \gamma_n c_{proc.}\left(q_{c1.n} + q_{c2.n}\right) \\ & - c_{inv.}\left[I_{n-1} - q_{p.n} + \gamma_n\left(q_{c1.n} + q_{c2.n}\right)\right] - \left[c_f + c_v \cdot \min\left(q_{p.n}, I_{n-1}\right)\right]\Big\}\end{aligned} \tag{12}$$

ls_1 : Lump-sum payment to supplier 1.
ls_2 : Lump-sum payment to supplier 2.

The contract structure shown in **Figure 7** with an adjustable range for procurement could be considered a robust method, however, the lump-sum payment makes the manufacturer cost extra expenses when the required volume of components is less than the reference volume or no procurement disruptions occur. The determination of procurement volume by lump-sum payments makes the price competitiveness of the manufacturer lower. To improve the problem, a structure of 2-stage procurement cost without lump-sum payments is proposed in this study. **Figure 8** shows the concept of the contract structure.

Instead of lump-sum payments, the manufacturer should procure components at a higher unit price than the normal procurement price when the procurement volume is

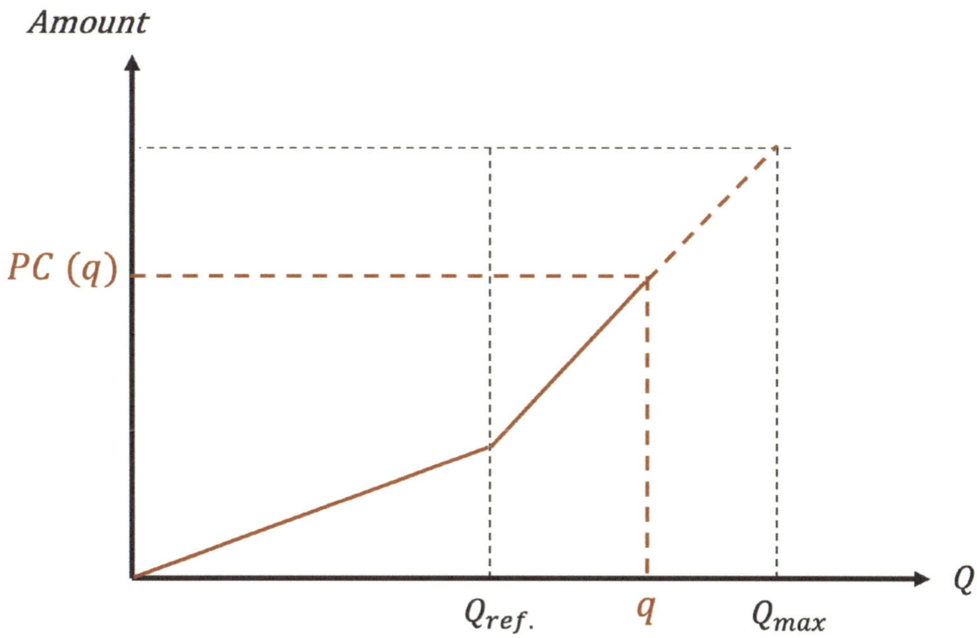

Figure 8.
Structure of 2-stage procurement cost.

more than the reference procurement volume according to the result of procurement allocation. $PC(q)$ is the total value of $Q_{ref.}$ with normal procurement price and procurement price to the additional volume of $(q - Q_{ref.})$. Moreover, the upper limit of procurement volume is determined by the supplier considering the unit price of the additional procurement volume. The higher the price of the additional procurement volume is the incentive of the supplier for setting a higher upper limit would be higher.

Eq. (13) shows the profit of the manufacturer with this contract structure while $q_{c.n} \leq Q_{ref.}$. Eq. (14) shows the profit of the manufacturer with this contract structure while $Q_{ref.} < q_{c.n} \leq Q_{max}$.

$$\pi_m = \sum_{n=1}^{N} \Big\{ p_p \cdot \min\left(q_{p.n}, I_{n-1}\right) - \gamma_n c_{proc.} \left(q_{c1.n} + q_{c2.n}\right) - c_{inv.}\left[I_{n-1} - q_{p.n} + \gamma_n\left(q_{c1.n} + q_{c2.n}\right)\right] - \left[c_f + c_v \cdot \min\left(q_{p.n}, I_{n-1}\right)\right]\Big\} \tag{13}$$

$$\pi_m = \sum_{n=1}^{N} \Big\{ p_p \cdot \min\left(q_{p.n}, I_{n-1}\right) - \gamma_n c_{proc.} \left(Q_{ref.1} + Q_{ref.2}\right) - \gamma_n c'_{proc.}\left(q_{c1.n} + q_{c2.n} - Q_{ref.1} - Q_{ref.2}\right) - c_{inv.}\left[I_{n-1} - q_{p.n} + \gamma_n\left(q_{c1.n} + q_{c2.n}\right)\right] - \left[c_f + c_v \cdot \min\left(q_{p.n}, I_{n-1}\right)\right]\Big\} \tag{14}$$

$Q_{ref.1}$: Reference procurement volume of supplier 1.
$Q_{ref.2}$: Reference procurement volume of supplier 2.
$c_{proc.}$: Normal component procurement unit cost.
$c'_{proc.}$: Component procurement unit cost for additional volume

It is considered that the competitiveness of the contract structure which is shown in **Figure 7** should be higher than the contract structure which is shown in **Figure 8** when product demand is in an expanding trend with high fluctuation. On the other hand, when product demand is in a stable trend with low fluctuation, it is considered that the competitiveness of the contract structure which is shown in **Figure 8** should be higher than the contract structure which is shown in **Figure 7**.

As shown in **Figure 9**, it is a combination structure of lump-sum payments and 2-stage procurement cost. The amount of lump-sum payments and procurement price to the additional volume are adjusted by the manufacturer with considering the features of product demand. The supplier should guarantee its supply volume by or over Q_{min} according to the amounts of lump-sum payments. It helps manufacturers to expand the range of procurement volume that can be procured reliably. Moreover, the 2-stage procurement cost gives the supplier the incentive to set its maximum component supply volume.

Profit of manufacturer with this contract structure could be expressed by Eq. (12) while $q_{c.n} \le Q_{min.}$. Eq. (15) shows the profit of the manufacturer with this contract structure while $Q_{min} < q_{c.n} \le Q_{max}$.

$$\pi_m = \sum_{n=1}^{N} \Big\{ p_p \cdot \min\left(q_{p.n}, I_{n-1}\right) - (ls_1 + ls_2) - \gamma_n c_{proc.} \left(Q_{min.1} + Q_{min.2}\right) \\ - \gamma_n c'_{proc.} \left(q_{c1.n} + q_{c2.n} - Q_{min.1} - Q_{min.2}\right) - c_{inv.}\left[I_{n-1} - q_{p.n} + \gamma_n \left(q_{c1.n} + q_{c2.n}\right)\right] \\ - \left[c_f + c_v \cdot \min\left(q_{p.n}, I_{n-1}\right)\right] \Big\} \tag{15}$$

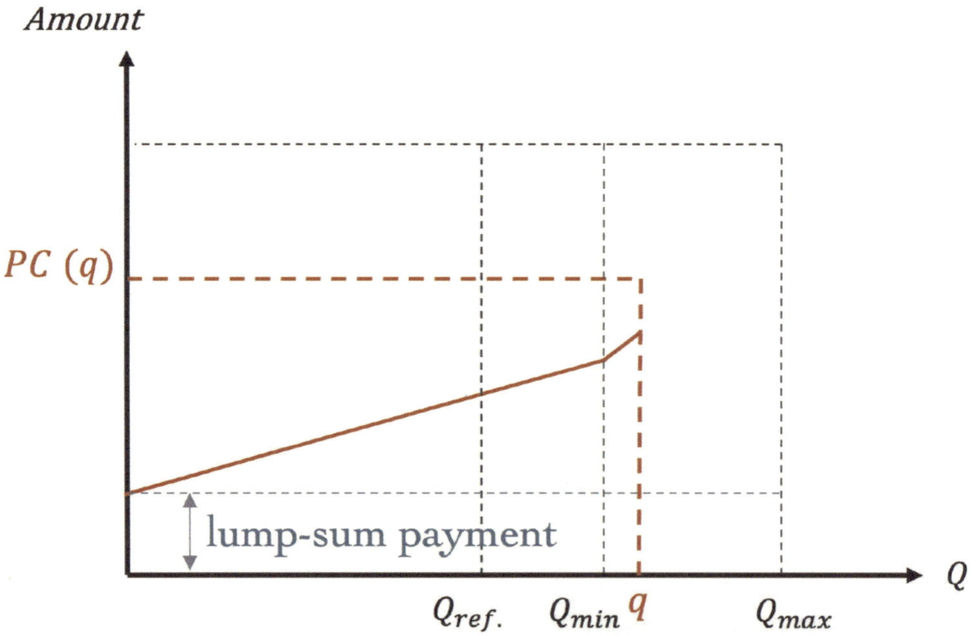

Figure 9.
Structure of lump-sum payment and 2-stage procurement cost.

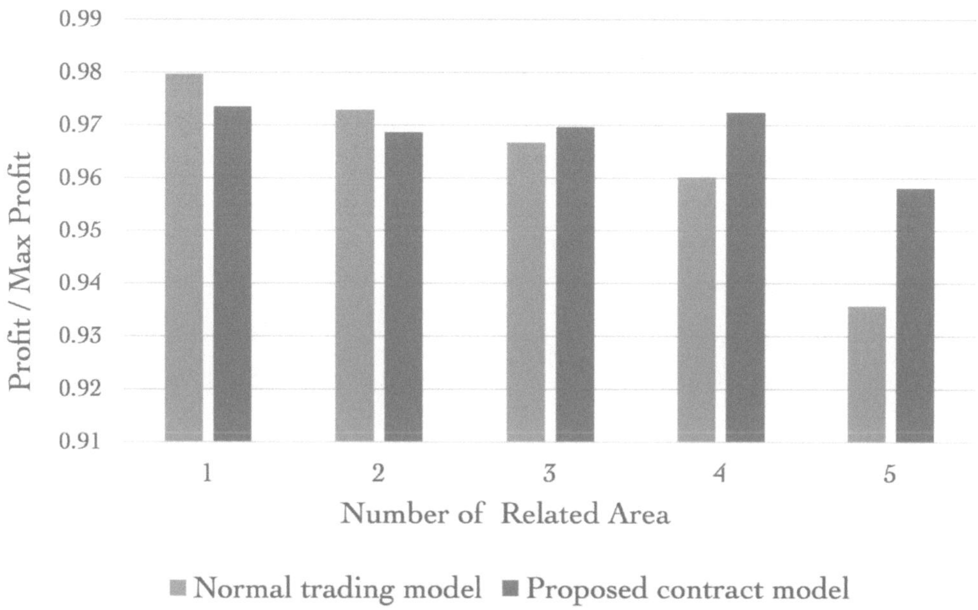

Figure 10.
Total profit of manufacturer by different related areas with normal trading model and proposed contract model.

4.2 Numerical experiment and results

The experiment environment is set as the same as in Section 2.3. It is considered that there are no differences in lump-sum payment, procurement volume and Q_{min} between two suppliers. Besides the experiment data in Section 2.3, Q_{min} is set to 10% more than the component order lot size, lump-sum payment is set as the component inventory cost of the difference between Q_{min} and order lot size. **Figure 10** shows an example of a numerical experiment result.

The vertical and horizontal axes of **Figure 10** are set as the same as in **Figure 5**. The results in **Figure 10** are also the average value of 50 simulation runs. The normal trading model is the model in which the manufacturer places an order to the supplier based on the estimated product demand, and the supplier supplies components according to the order at a pre-determined trading price. On the other hand, the proposed contract model is the contract structure of lump-sum payment and 2-stage procurement cost. From **Figure 10**, it could be found that with the proposed contract model, there is no significant reduction in profit with increasing in related areas.

5. Conclusions

As discussed in Section 1, many manufacturers are stuck between step 2 and step 3 of the BDP development and operation cycle. Lack of visibility of benefits under normal situations as well as mitigation effects against losses due to supply chain disruptions is considered the main reasons. **Figure 11** shows the detail of **Figure 2** which is the summary of prior measures discussion and design. In this chapter, except for the process of execution based on contract terms, the hierarchy of prior measures

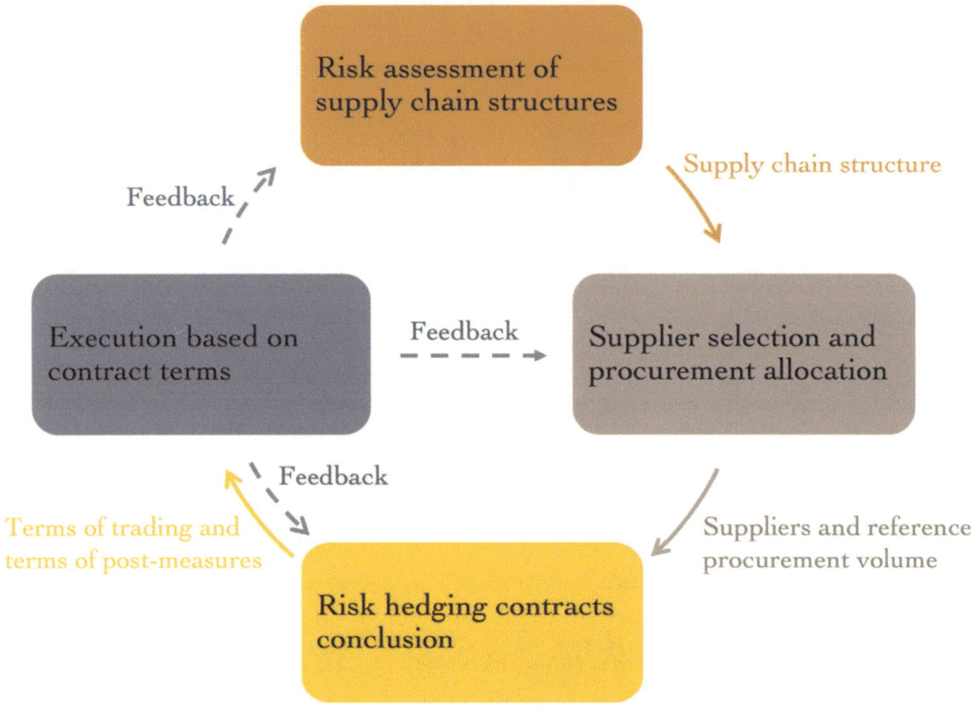

Figure 11.
Summary of prior measures discussion and design.

and ideas on specific measures, including risk assessment of supply-chain structures, supplier selection, procurement allocation, and risk-hedging contracts conclusion, is proposed. The effectiveness of these proposed measures is confirmed by numerical experiments.

In Section 2, the profit of manufacturers with a centralized procurement structure model and decentralized procurement structure model is confirmed under supply-chain disruption risks due to earthquakes in Japan. The loss aversion effect of a decentralized procurement structure is confirmed quantitatively. In Section 3, a decision-making method for a decentralized procurement structure is proposed based on Tabu Search. It is confirmed that it could help decision-makers to make more accurate solutions effectively in supplier selection and procurement allocation. In Section 4, the lump-sum payment and 2-stage procurement cost structure contract model based on the selected suppliers and the results of procurement allocation is proposed. It could help manufacturers to solve the problem of securing procurement volume while falling in procurement disruptions and the problem of keeping procurement at low cost even in normal periods.

Market power and cooperation level between upstream and downstream companies are necessary to be considered in designing and planning supply chains. Furthermore, all of the analyses and proposals in this chapter are from the perspective of the manufacturer. Actually, a robust and competitive supply chain could not be built and kept without cooperation from suppliers. Therefore, analysis from the perspective of the supplier to clarify the conditions for realizing a win–win relationship is also an important task.

Nomenclature list

p_p	Product price
$q_{p.n}$	Volume of product orders from the retailer
I_n	Inventory level of components
$c_{proc.}$	Component procurement cost/per unit
$q_{c.n}$	Volume of component procurement
$c_{inv.}$	Inventory cost/per unit
c_f	Fixed cost of production
c_v	Variable cost of production
N	Period (Month)
$q_{p.n}$	Volume of product orders from the retailer
$q_{c1.n}$	Volume of component procurement from supplier 1
$q_{c2.n}$	Volume of component procurement from supplier 2
β_n	If there is no procurement disruption, $\beta_n = 1$. Otherwise $\beta_n = 0$
γ_n	If there is no procurement disruption, $\gamma_n = 1$. Otherwise $\gamma_n = k$, If component procurement from supplier 2 is disrupted $k = q_{c1.n}/q_{p.n}$, and if component procurement from supplier 1 is disrupted $k = q_{c2.n}/q_{p.n}$. Otherwise, $k = 0$ since the probability of such a case is extremely low, it is not considered in the experiment of this study.
n	Number of suppliers
ZX_s	If components are sourced from the supplier s, $ZX_s = 1$. Otherwise $ZX_s = 0$.
Q_{is}	Procurement volumes of components i from supplier s
s	Supplier ID ($s = 1, 2, 3, \dots, S$)
i	ID of component types ($i = 1, 2, 3, \dots, I$)
D_i	Demand for component i
M_s	Management cost of supplier s
P_{is}	Price of component i from supplier s (per unit)
Q_{is}^{min}	Minimum procurement volume of components i from supplier s
Q_{is}^{max}	Maximum procurement volume of components i from supplier s
L_i	Opportunity loss of components i (per unit)
$Trans_{is}$	Transportation cost of components i from supplier s (per unit)
θ_s	Disruption probability of supplier s (%)
a	ID of interrupted suppliers among selected suppliers ($a = 1, 2, 3, \dots, A$)
b	ID of among selected suppliers ($b = 1, 2, 3, \dots, B$)
A_n	The interrupted supplier group (Ex. A_1 indicates the group of anyone interrupted supplier)
B_n	The uninterrupted supplier group (Ex. B_1 indicates the group of uninterrupted suppliers when any supplier is interrupted)
Q_{ib}^{max}	Maximum procurement volume of components i from uninterrupted suppliers
C_n	The amount of opportunity loss due to any n procurement channels getting disrupted
Q_{max}	Upper limit of procurement volume determined by the supplier according to the unit price of the additional procurement volume
ls_1	Lump-sum payment to supplier 1
ls_2	Lump-sum payment to supplier 2

p_p	Product price
$q_{p.n}$	Volume of product orders from the retailer
I_n	Inventory level of components
$c_{proc.}$	Component procurement cost/per unit
$c_{inv.}$	Inventory cost/per unit
c_f	Fixed cost of production
c_v	Variable cost of production
N	Period (Month)
$Q_{ref.1}$	Reference procurement volume of supplier 1
$Q_{ref.2}$	Reference procurement volume of supplier 2
$c'_{proc.}$	Component procurement unit cost for additional volume
$q_{c.n}$	Volume of component procurement
$q_{c1.n}$	Volume of component procurement from supplier 1
$q_{c2.n}$	Volume of component procurement from supplier 2
$Q_{min\ .1}$	Upper limit of procurement volume from supplier 1 according to the amounts of lump-sum payments to supplier 1
$Q_{min\ .2}$	Upper limit of procurement volume from supplier 2 according to the amounts of lump-sum payments to supplier 2
Q_{max}	Upper limit of procurement volume determined by the supplier according to the unit price of the additional procurement volume
γ_n	If there is no procurement disruption, $\gamma_n = 1$. Otherwise $\gamma_n = k$, If component procurement from supplier 2 is disrupted $k = q_{c1.n}/q_{p.n}$, and if component procurement from supplier 1 is disrupted $k = q_{c2.n}/q_{p.n}$. Otherwise, $k = 0$ since the probability of such a case is extremely low, it is not considered in the experiment of this study.

Author details

Kotomichi Matsuno[1*], Jiahua Weng[2], Noriyuki Hosokawa[1] and Takahiro Ohno[1]

1 Waseda University, Tokyo, Japan

2 Kanagawa University, Kanagawa, Japan

*Address all correspondence to: sidiwu@aoni.waseda.jp

References

[1] Business Policy Forum. Survey Report on Improving the Effectiveness of Corporate Business Continuity in the Wake of the Great East Japan Earthquake [Internet]. 2013. Available from: https://www.bpfj.jp/report/manufacturing_h24/ [Accessed: 2022-02-25]

[2] The Small and Medium Enterprise Agency. Guidebook on SME Business Continuity Planning [Internet]. 2008. Available from: https://www.chusho.meti.go.jp/keiei/antei/download/bcp_guide.pdf . [Accessed: 2022-02-25]

[3] Maruya H, zigyou keizokukeikaku no igitokeizaikouka: heizyouzi ni hyoukasareru zissenmanezimenntohe (in Japanese). Tankobon. Gyosei; 2008. p. 244

[4] Matsuno K, Hosokawa N, Ohno T. Business analysis of electronic device manufacturers on business continuity plans under uncertain supply chain disruption risks. In: E-Proceedings of the 26th International Conference on Production Research (26th ICPR). Taichung; 18-21 July 2021

[5] Tokui J, Kawasaki K, Miyagawa T. The economic impact of supply chain disruptions from the great East-Japan earthquake. Japan and the World Economy. 2017;**41**:59-70. DOI: 10.1016/j.japwor.2016.12.005

[6] Park Y, Hong P, Roh JJ. Supply chain lessons from the catastrophic natural disaster in Japan. Business Horizons. 2013;**56**:75-85. DOI: 10.1016/j.bushor.2012.09.008

[7] Masuda T, Mizuno H. Study for risk management of disruption in supply chains. Proceedings of the School of Information and Telecommunication Engineering Tokai University. 2017; **10**(1):25-35

[8] Matsukawa H, Liu T. A supply chain visualization system and its validation analysis. Journal of Japan Industrial Management Association. 2016;**67**(2): 73-82. DOI: https://doi.org/10.11221/jima.67.73

[9] Japan Weather Association: Historical Information of Earthquakes [Internet]. Available from: https://earthquake.tenki.jp/bousai/earthquake/entries/level-6-plus/. [Accessed: 2022-02-27]

[10] Kashima H. The 2016 Kumamoto earthquake and manufacturing industry: Regional characteristics on the disasters and restoration processes. Annals of the Association of Economic Geographers. 2018;**64**(2):138-149. DOI: https://doi.org/10.20592/jaeg.64.2_138

[11] Matsuno K, Weng J, Shao X. Sourcing decision with capacity reservation under supply disruption risk. Asian Journal of Management Science and Applications. 2021;**6**(1):49-68. DOI: 10.1504/AJMSA.2021.118403

[12] Moghaddam KS. Fuzzy multi-objective model for supplier selection and order allocation in reverse logistics systems under supply and demand uncertainty. Expert Systems with Applications. 2015;**42**(15–16):6237-6254. DOI: 10.1016/j.eswa.2015.02.010

[13] Arabsheybani A, Paydar MM, Safaei SA. An integrated fuzzy MOORA method and FMEA technique for sustainable supplier selection considering quantity discounts and supplier's risk. Journal of Cleaner Production. 2018;**190**:577-591. DOI: 10.1016/j.jclepro.2018.04.167

[14] Abginehchi S, Farahani Z. R: Modelling and analysis for determining optimal suppliers under stochastic lead

times. Applied Mathematical Modelling. 2010;**34**(5):1311-1328. DOI: 10.1016/j.apm.2009.08.021

[15] Meena P, Sarmah PS, Sarkar A. Sourcing decisions under risks of catastrophic event disruptions. Transportation Research Part E: Logistics and Transportation Review. 2011;**47**(6): 1058-1074. DOI: 10.1016/j.tre.2011.03.003

[16] Chopra S, Reinhardt G, Mohan U. The importance of decoupling recurrent and disruption risks in a supply chain. Naval Research Logistics. 2007;**54**(5): 544-555. DOI: 10.1002/nav.20228

[17] Shu T, Yang F, Chen S, Wang S, Lai KK, Gan L. Contract coordination in dual sourcing supply chain under supply disruption risk. Mathematical Problems in Engineering. 2015;**2015**(7):1-10. DOI: 10.1155/2015/473212

[18] Asian S, Nie X. Coordination in supply chains with uncertain demand and disruption risks: Existence, analysis, and insights. IEEE Transactions on Systems, Man, and Cybernetics: Systems. 2014;**44**(9):1139-1154. DOI: 10.1109/TSMC.2014.2313121

[19] Hou J, Zeng AZ, Sun L. Backup sourcing with capacity reservation under uncertain disruption risk and minimum order quantity. Computers and Industrial Engineering. 2017;**103**: 216-226. DOI: 10.1016/j.cie.2016.11.011

Chapter 3

Facility Location

Nneoma Benita Amos and Edafe Bawa Dogo

Abstract

The location of the production, distribution and storage facilities of organizations, remains one of the most important strategic decision production and operations managers have to make. Minimizing total cost and distance traveled is the fundamental objective in deciding the location of facilities in order to maximize total profit. There are several quantitative methods for solving facility location problems and deciding on the most optimal decision to reach as it regards locating a facility or assigning workers to their most optimal location or tasks. The Linear Assignment model with particular emphasis on the Hungarian and Branch and Bound technique under minimization and maximization situations are deterministic approaches to solving facility location problems and have been discussed in this chapter.

Keywords: facility location, Hungarian, branch and bound, linear assignment model, optimality

1. Introduction

Facility Location is an important factor in the supply chain that significantly impacts on the efficiency and effectiveness of many supply networks and the organization at large. Location decisions are strategic in character, long-term in nature, and non-repetitive in nature. Without good and thorough site planning from the start, the new facilities may have ongoing operational issues in the future. Poor Location decision not only affects the growth of the firm but impedes on the growth and development of the nation. The location decision should be made with great care, as any error that results in a poor location can be a constant source of higher costs, higher investment, difficult marketing and transportation, dissatisfied and frustrated employees and consumers, frequent interruptions of production, abnormal wastages, delays, and substandard quality, among other things.

Facility location significantly impacts on revenue, costs, and service levels to customers. It is thus a classical optimization problem for determining the sites for factories, service outlets and warehouses. Facility location decision is made by selecting the best option among a set of possible sites depending on the nature or type of business. The choice of facility location is strategically guided by profit maximization or minimization of all costs associated with the choice of location.

2. Facility location planning

Facility location is connected with capacity decisions. Capacity expansion considerations instantly raises the twin issue of where to expand in order to tie in effectively with the distribution network of facility location. Facility location of operations is a long-term capacity decision which involves huge and long term commitment about the geographically fixed factors that affects business organizations. The selection of location is therefore a key-decision of production and operations managers as large investment is made in building plant and machinery.

Cambridge dictionary defined a facility as "a place, especially including buildings, where a particular activity happens". Facility extends beyond a place or a building, it includes structures, equipment, or people. It also includes hospitals, food production plants, and gas stations among others. Adeleke and Olukanmi [1] suggested that facility location issues seeks to determine how to locate a number of facilities from a set of potential facilities that will serve a number of customers. Alenezy [2] added that the cost-effective site is to be chosen from the potential locations in which to place new facilities or retain existing ones.

Facility location can therefore be defined as the siting of facilities which could be structures, men, material, machines in such a manner that yields an optimum benefit to the firm and its stakeholders. There are two major location decisions or problems faced by managers, they are internal and external location problems. Internal location problems are concerned with decisions on where to locate facilities inside the plant. A good example of an internal location decision is where to place a new machine or storage room within the existing facility. Internal facility location decision deals with the assignment or location of facility whose space constraint is equal to or smaller than the available space existing within the facility. An external location decision on the other hand deals with the problem of where to site a new manufacturing plant, warehouse, building, or a new branch within an agreed geographical area [3–7].

The following circumstances may necessitate facility location decisions:

1. Upon the commencement of a new business or gaining entrance

2. When an established company outgrows its existing facility and expansion is no longer feasible, a new location decision is made.

3. Expansion of current firm or market size that demands the opening of new branches.

4. Insecurity issues

5. Creation/development of a new product.

6. Government policies.

7. When a lease ends without being renewed by the landlord.

8. Social or financial considerations

When expansion is inevitable three options are open to an organization

- Expand on the current location of land/facilities available
- Look for a new space and use it as an extension of the other office
- Look for a large portion of land that can accommodate both the existing business and the new line of business

3. Importance of location decision

i. Location study helps to find the optimal location of organization facility or plant that will result in the greatest advantage to the organization.

ii. Location plays a huge role in attracting and retaining the best employees, that gives a competitive edge over the firm's rival

iii. A good location decision helps in avoiding waste of all the investments made in plant and machinery equipment.

iv. A good location decision helps to optimize performance of the firm by minimizing its total cost of production.

v. It also ensures safety for its workforce

vi. Locating a facility in the right place can give access to customers and also enable the firm enjoy incentives provided by government

4. Factors to consider when taking a location decision

There are several factors that must be critically analyzed when considering locating a facility. These factors can be grouped into two categories: controllable and uncontrollable factors.

4.1 Controllable factors

i. **Availability of Inputs:** This is a major decision guiding facility location problems. Manufacturing companies that use heavy, bulky or perishable products as raw materials or factor inputs have to be located near the source of these raw materials or inputs, this is to ensure regular and timely supply of raw materials as well as reduce the cost of transporting and storing them. A good example are food processing companies, the major farm product used in processing those goods need to be in close range to ensure steady supply of these products. Similarly, most wood processing companies are located close to the supply of quality timber. Generally, the cost of transportation or shipping of these raw material is to be weighed. Perishable products may be lost or damaged in transit if the distance to the plant is far, thus the closer the facility is to the source of its inputs determines its ability to cut costs.

ii. **Market/Customer Proximity:** Goods are produces to be sold to the identified markets or customers, therefore the proximity of a facility to the market or customers is of grave importance to the organization. Proximity helps to reduce transportation costs and time of delivery.

iii. **Integration with Other Parts of the Organization:** It is of significant benefit to keep a new plant or subsidiary plant close to the parent facility. This makes it easier for them to share resources and thus reduce total cost.

iv. **Availability of Labor and Skills:** Education, experience and skill of available labour is an important factor that affects location decision. It is always preferable to locate the plant in an area where skilled, semi-skilled and unskilled labour are available. This reduces costs of training and hiring experts from abroad

v. **Availability of Amenities:** Good roads, hospitals, school's churches, parks and residential area are necessary amenities that make living conditions for worker's desirable.

vi. **Availability of Transportation Facilities:** Good transportation facilities makes the plants accessible for easy movement of raw materials as well as finished products, thereby reducing costs.

vii. **Availability of Services:** The availability of basic support services needed by the facility to make their operations smooth should be considered. This decision affects to total cost of operation, where such services are not readily available or at a high cost, the organization will have to spend more in providing such services, or attracting them at a higher cost.

viii. **Room for Expansion:** The possibility of increasing future production capacity occasioned by increase in product demand is a critical factor in location decisions. There should be adequate space for future expansion or diversification of the facility as the need arises.

ix. **Safety:** The safety of employees as well as the facility needs to be taken into consideration when making location decisions. If the location is not safe, it may detract employees and even potential customers from patronizing the facility. Also, the probability of loss of property, or damage of the machines increases when safety of the environment or location is poor.

4.2 Uncontrollable factors

i. **Community and labour attitude:** Communities that are interested in attracting new plants may offer reduced prices or no cost sites to companies as a way of growing their communities. The cost of land and attitude of labour to work is a major factor to consider when making location decisions.

ii. **Suitability of Land and Climate:** Due consideration should be given to the suitability of the land and climate for the nature of the products and the type of machines used in production. If the climate in a given geographical location does not support the product type it may lead to wastage in the long run, hence due consideration should be given to this factor.

iii. **Regional Regulations:** The regulations in certain regions do not support the production of certain types of products or services, organizations should investigate the regulations of the desired region before going ahead to site their plants or facility.

iv. **Political, Cultural and Economic Situation:** The political, cultural and economic situation of the location should be well considered. Areas notable for political unrest, may not favor the facility as protests and other activities may lead to the damage of the facility. Also, some cultures are noted for certain behaviors that may not be compatible with the activities of the plant. Lastly, economic situation in a given location may stall the growth of the organization in the short and long run.

v. **Power supply:** Cost and quality of power supply is a high importance in the location of a plant. Cost of power supply is usually cheaper at rural locations than the urban areas. Some companies generate their own power. The overall cost of power supply should be taken into consideration when making facility location decisions.

vi. **Regional Taxes, Special Grants and Import/Export Barrier:** The kind and amount of taxes levied by a state should be considered in locating a plant/facility. Investigation should be made on the type of taxes and the biases for which they are fixed. Some places have special grants given to attract investors to such area. Similarly, import/export barriers should be duly considered as this affects the total cost of doing business.

vii. **Government Policies:** The policies of the state governments and local bodies concerning labor laws, building codes, safety, etc., are the factors that demand attention. In order to have a balanced regional growth of industries, both central and state governments in our country offer the package of incentives to entrepreneurs in particular locations. The incentive package may be in the form of exemption from a sales tax and excise duties for a specific period, soft loan from financial institutions, subsidy in electricity charges and investment subsidy. Some of these incentives may tempt to locate the plant to avail these facilities offered.

5. Solution methods for location problem

There are various approaches for solving location problems such as:

i. The Dimensional Analysis approach

ii. The Simple ranking of alternative site approach

iii. The Linear Assignment model

iv. The Linear Transportation model

v. The Quadratic Assignment model

vi. The Heuristic solution procedure

The solution method we shall focus on is the Linear Assignment model.

6. Linear assignment model

Linear assignment model (LAM) is a deterministic mathematical model applied when the management decides to make an optimum use of its capacity. It is referred to as being linear because there is always a straight line relationship between the variable involved. There are two major objectives which linear assignment model can seek to achieve;

i. Maximization of profit

ii. Minimization of cost

There are two major methods for solving linear assignment model

1. Hungarian method

2. Branch and Bound technique

6.1 Hungarian method

The Hungarian method is one of the many algorithms that have been devised to solve linear assignment problem.

To solve identified problem, the number of assignment must be equal to the number of people or machines to do the job. However, if there is a shortfall in either in the available job or persons/machines to do the job. A dummy row or column is created before carrying out the various iterations to solve the problem.

6.1.1 Procedures for solving minimization problems

When dealing with minimization objective under the Hungarian method, it is critical to ensure that the scenario in the case is a minimization situation. If the question focusses on cost, risk, distance covered etcetera then it is a minimization question and the following ten simple steps can be applied.

Step 1: Write out the initial matrix and check whether the number of row is equal to the number of columns. Create a dummy row or column where necessary.

Step 2: Perform row reduction by identifying the smallest element in each row and deduct the identified element from the other element in the row. Do same for all the rows.

Step 3: Perform a column reduction by identifying the smallest element in each column and deducting it from all other element. Apply to all columns. Please note this is to be done on the most recent matrix after the row reduction and not on the initial matrix.

Step 4: identify the number of unique zero's in each selected row and column.

Step 5: Apply the minimum number of possible lines on the unique zero's starting with either the row or column with the highest number of unique zero's. Where there are ties, they are broken arbitrarily. Note: do not cover any unique zero twice.

Step 6: Count the number of minimum possible lines and compare with the number of assignment to be performed i.e. M = N = Minimum number of lines ruled (//).

Where M = Persons or machines to be assigned the task.

N = Assigned task.

Step 7: If M = N= // (optimality had been attained). Jobs can then be assigned to workstations or locations but if M $\neq$ N $\neq$//or if M = N $\neq$// (optimality is not attained). The iteration process should be continued.

Step 8: If optimality is not attained then identify the smallest uncovered element in the matrix and deduct same from other uncovered elements in the matrix, but add the identified smallest uncovered element to the elements at the point of intersection of the straight lines.

Step 9: Repeat steps 4, 5, 6 above until optimality is attained (i.e M = N= //).

Step 10: Assign jobs starting with the row with a single unique zero and pick the corresponding value in the initial matrix.

Example 1

A Plant manager has four subordinates and four tasks to be performed. The subordinates differ in efficiency and the tasks differ in their intrinsic difficulty. The estimate of the time each man would take to perform each task is given in the effectiveness matrix below. The objective is to assign men to jobs in such a way that the total time taken to complete an assignment is minimized (see **Table 1**).

Subordinates	T1	T2	T3	T4
Sands	8	26	17	11
Armstrong	13	28	4	26
Luke	38	19	18	15
Waters	19	26	24	10

Table 1.
Cost matrix.

Solution:
(see **Tables 2–5**).

Subordinates	T1	T2	T3	T4	Smallest element in a row
Sands	8	26	17	11	**8**
Armstrong	13	28	4	26	**4**
Luke	38	19	18	15	**15**
Waters	19	26	24	10	**10**

Table 2.
Initial tableau.

Subordinates	T1	T2	T3	T4
Sands	0	18	9	3
Armstrong	9	24	0	22
Luke	23	4	3	0
Waters	9	16	14	0
Smallest element in a column	**0**	**4**	**0**	**0**

Table 3.
Row reduction matrix.

Subordinates	T1	T2	T3	T4
Sands	0	14	9	3
Armstrong	9	20	0	22
Luke	23	0	3	0
Waters	9	12	14	0

Table 4.
Column reduction matrix.

SUBORDINATES	T1	T2	T3	T4
SANDS	0	14	9	3
ARMSTRONG	9	20	0	22
LUKE	23	0	3	0
WATERS	9	12	14	0

Table 5.
Straight line tableau.

Test for optimality
M = N = No of lines
4 = 4 = 4

From the table above, all the zeros in the rows and columns have been covered by four straight lines. Hence, the optimal solution has been reached and we can now make an optimal assignment following the pattern of unique zeros (see **Table 6**).

Subordinates	Possible tasks	Assigned tasks	Time
Sands	T1	T1	8
Armstrong	T3	T3	4
Luke	T2; T4	T2	19
Waters	T4	T4	10
			41 HOURS (OPTIMAL)

Table 6.
Assignment table.

The assignment should begin with the subordinate with only one possible task based on the number of unique zeros on each row. From the straight line table we can see that Sands, Armstrong and Waters have one unique zero on their respective rows, hence we assign Sands to Task 1, Armstrong to T3 and Waters to T4. Luke has two unique zeros on the row indicating two possible assignments (T2 and T4) since T4 has already been assigned to Waters (being the only possible assignment to Waters) we can no longer assign it to another subordinate, we then assign T2 to Luke. The summary of our assignment is shown above.

Example 2

The management of Cadbury Plc is interested in assigning her newly employed operations managers to her newly established branches in four different locations in Africa (Nigeria, Ghana, South Africa and Kenya). The cost implication and profit are of interest to the firm. However, the management desires to focus on cost (see **Table 7**).

Managers/Location	Nigeria	Ghana	SA	Kenya
Mike	950	610	980	360
Sultan	960	760	890	1040
Emmanuel	600	940	670	850
Fisher	750	650	1140	750

Table 7.
Allocation cost per month.

Assist the management to assign the newly employed managers to an appropriate location in the most optimal order and calculate the total cost of the operations per quarter using the information in the matrix above.

Solution: From the above Number of location = No of directors so no need for a dummy.

(See **Tables 8–10**).

Managers/Location	Nigeria	Ghana	SA	Kenya
Mike	950	610	980	360
Sultan	960	760	890	1040
Emmanuel	600	940	670	850
Fisher	750	650	1140	750

Table 8.
Initial matrix.

Managers/Location	Nigeria	Ghana	SA	Kenya	Smallest element in each row
Mike	950	610	980	360	360
Sultan	960	760	890	1040	760
Emmanuel	600	940	670	850	600
Fisher	750	650	1140	750	650

Table 9.
Row reduction table.

Managers/Location	Nigeria	Ghana	SA	Kenya
Mike	590	250	550	0
Sultan	200	0	60	280
Emmanuel	0	340	0	250
Fisher	100	0	420	100

Table 10.
Column reduction table.

Identify the smallest element on the first column and deduct that value from other values in the column, do this for all the columns (see **Table 11**).

Managers/Location	Nigeria	Ghana	SA	Kenya	Number of unique zero's in each row
Mike	590	250	550	0	1
Sultan	200	0	**60**	280	1
Emmanuel	0	340	0	250	2
Fisher	100	0	420	100	1
Number of unique zero's in each column	1	2	1	1	

Table 11.
Applying the straight line rule.

Identify the number of unique zeros in each row and column and start applying straight lines across the zeros starting from the column or row with the highest number of zeros until all the unique zeros are covered.

From the above table the number of rows is equal to the number of columns but not equal to the number of straight lines that is M = N ≠ // Optimality has not been attained.

We then proceed to identify the smallest uncovered element from the table.

The smallest uncovered element = 60.

Subtract 60 from all other uncovered elements but add it to the values in the cells at the point of intersection this will produce a new matrix table (see **Tables 12** and **13**).

M = N = // (Number of lines ruled) Optimality has been attained because the number of minimum possible lines is equal the assigned task (see **Table 13**).

The optimal assignment table (**Table 14**) shows that following the rule of assigning we have a total cost per month of 2,500,000. To obtain the total cost per quarter multiply the cost per month by three months that make up a quarter as shown below.

$$\text{Total Cost per quarter} = \text{N}2,500,000\text{X}3 = \text{N}7,500,000.$$

Managers/Location	Nigeria	Ghana	SA	Kenya
Mike	590	310	550	0
Sultan	140	0	0	220
Emmanuel	0	400	0	250
Fisher	40	0	360	40

Table 12.
New matrix.

Managers/Location	Nigeria	Ghana	SA	Kenya	Number of unique zero's in each row
Mike	590	310	550	0	1
Sultan	140	0	0	220	2
Emmanuel	0	400	0	250	2
Fisher	40	0	360	40	1
Number of unique zero's in each column	1	2	2	1	

Table 13.
Applying the straight line rules.

Managers	Possible Locations	Assigned Location	Cost (N)
Mike	Kenya	Kenya	360,000
Sultan	Ghana /SA	SA	650,000
Emmanuel	Nigeria/ SA	Nigeria	890,000
Fisher	Ghana	Ghana	600,000
			2,500,000

Table 14.
Optimal assignment table.

Example 3

A company wishing to supply its products to a region with five major areas has five trucks to accomplish this task. The matric table below shows the time in hours it will take each of these trucks to service an area. Assuming a truck can service only one area, make an assignment of trucks to areas in a way that will minimize available time (see **Table 15**).

Trucks\Area	I	II	III	IV	V
A	8	32	24	17	50
B	12	42	32	23	60
C	10	30	15	9	30
D	50	60	10	11	15
E	24	40	50	16	17

Table 15.
Truck time matrix.

Solution.
Step I:
(See **Table 16**).
Step II:
(See **Table 17**).
Step III:
(See **Table 18**).
Step VI: identify the number of unique zeros in each row and column.
Step V.
(See **Table 19**).

Trucks\Area	I	II	III	IV	V	Smallest element on each row
A	8	32	24	17	50	8
B	12	42	32	23	60	12
C	10	30	15	9	30	9
D	50	60	10	11	15	10
E	24	40	50	16	17	16

Table 16.
Initial matrix.

Trucks\Area	I	II	III	IV	V
A	0	24	16	9	42
B	0	30	20	11	48
C	1	21	6	0	21
D	40	50	0	1	5
E	8	24	34	0	1
Smallest element on each column	0	21	0	0	1

Table 17.
Row reduction table.

Trucks\Area	I	II	III	IV	V
A	0	3	16	9	41
B	0	9	20	11	47
C	1	0	6	0	20
D	40	29	0	1	4
E	8	3	34	0	0

Table 18.
Column reduction table.

Trucks\Area	I	II	III	IV	V
A	0	3	16	9	41
B	0	9	20	11	47
C	1	0	6	0	20
D	40	29	0	1	4
E	8	3	34	0	0

Table 19.
Straight line table.

Trucks\Area	I	II	III	IV	V
A	0	0	13	6	38
B	0	6	17	8	44
C	4	0	6	0	20
D	43	29	0	1	4
E	11	3	34	0	0

Table 20.
New matrix.

Trucks\Area	I	II	III	IV	V
A	0	0	13	6	38
B	0	6	17	8	44
C	4	0	6	0	20
D	43	29	0	1	4
E	11	3	34	0	0

Table 21.
Straight line table.

Step VI: Optimality Check.

M = 5, N = 5, Number of straight lines equals 4.

Therefore optimality has not been attained so we carry out step VII.

Step VII: the least uncovered element is 3, we subtract it from all uncovered elements and add it to the point of intersection, the new table we obtain is shown below.

(See **Table 20**).

Repeat from step IV to step VI.

Step IV: Identify the number of unique zeros in each row and column.

Step V:

(See **Table 21**.)

Step VI: Optimality Check.

M = 5, N = 5, Number of straight lines equals 5.

Therefore optimality has been attained since M = N = \\.

(See **Table 22**).

Advice

Trucks	Possible assignment	Assigned area	Cost
A	I, II	II	32
B	I	I	12
C	II, IV	IV	9
D	III	III	10
E	IV, V	V	17
			80

Table 22.
Assignment table.

From the assignment table above truck A should be assigned to area II, truck B to area I, C to IV, D to III and truck E to area V in this case the total time it will take the five trucks to complete the assignment will be 80 hours.

6.1.2 Procedures for solving maximization problem

There are two alternatives that can be used in solving a maximization problem. This labeled as Alternative A approach and Alternative B approach as a means of differentiating the two methods (see **Table 23**).

Steps	Alternative A	Alternative B
1	Write out the initial matrix	Write out the initial matrix
2	Identify the highest figure in the matrix and deduct other elements from it	Identify the highest element in each selected row and deduct other element in each row from it
3	Perform row reduction operation using minimisation methods	Perform column reduction operation using minimisation methods
4	Perform column reduction as explained in the minimisation procedure	
5	Repeat steps 4–8 as done in the minimisations objective procedure	Repeat steps 4–8 as done in the minimisations objective procedure

Table 23.
Procedures for solving maximization problem.

Example 1

First Bank Nigeria Plc has just completed a recruitment exercise and wants to assign her newly trained accountants to the location that will enhance the efficiency of the bank. Assuming that the matrix below represents the efficiency score of the accountants (see **Table 24**).

As an operation manager assign the newly trained accountants to the most optimal location and estimate the efficiency score.

Solution;

Using Alternative A Approach.

(See **Table 25**).

Convert the matrix values to relative cost by subtracting all values in the matrix from the highest values in the matrix.

(See **Tables 26–29**).

M = N = Number of straight lines ruled (optimality had been attained).

Accountants/Location	Location I	Location II	Location III	Location IV
Peter	225	120	255	180
Osha	245	100	190	315
Rose	105	215	170	175
Lamark	190	110	129	200

Table 24.
Efficiency matrix table.

Accountants/Location	Location I	Location II	Location III	Location IV
Peter	225	120	255	180
Osha	245	100	190	315 highest
Rose	105	215	170	175
Lamark	190	110	129	200

Table 25.
Initial matrix.

Accountants/Location	Location I	Location II	Location III	Location IV	Lowest in each row
Peter	90	195	60	135	60
Osha	70	215	125	0	0
Rose	210	100	145	140	100
Lamark	125	205	186	115	115

Table 26.
Conversion to cost matrix table.

Accountants/Location	Location I	Location II	Location III	Location IV
Peter	30	135	0	75
Osha	70	215	125	0
Rose	110	0	45	40
Lamark	10	90	71	0
Smallest in each column	10	0	0	0

Table 27.
Row reduction.

Accountants/Location	Location I	Location II	Location III	Location IV
Peter	20	135	0	75
Osha	60	215	125	0
Rose	100	0	45	40
Lamark	0	90	71	0

Table 28.
Column reduction.

(See **Table 30**).

The total efficiency score from the optimal assignment table above is 975.

Using Alternative B Approach

(See **Tables 31–34**).

M = N = Number of Straight lines ruled (optimality is attained).

Accountants/Location	Location I	Location II	Location III	Location IV	Number of Unique Zero's
Peter	20	135	0	75	1
Osha	60	215	125	0	1
Rose	100	0	45	40	1
Lamark	0	90	71	0	2
No of unique zero	1	1	1	2	2

Table 29.
Applying the straight line rules.

Accountant	Possible Branches	Assigned Branches	Efficiency Score
Peter	Location III	Location III	255
Osha	Location IV	Location IV	315
Rose	Location II	Location II	215
Lamark	Location I/Location IV	Location I	190
		Total Score	**975**

Table 30.
Optimal assignment table.

Location/ Accountants	Location I	Location II	Location III	Location IV	Highest element in each row
Peter	225	120	**255**	180	**255**
Osha	245	100	190	**315**	**315**
Rose	105	**215**	170	175	**215**
Lamark	190	110	129	**200**	**200**

Table 31.
Initial matrix.

Location/ Accountants	Location I	Location II	Location III	Location IV
Peter	30	135	0	75
Osha	70	215	125	0
Rose	110	0	45	40
Lamark	10	90	71	0
Least in Column	**10**	**0**	**0**	**0**

Table 32.
Row reduction.

(See **Table 35**).
The total efficiency score from the optimal assignment table above is 975.
NOTE: The same answer will be arrived at no matter the method adopted.

Location/ Accountants	Location I	Location II	Location III	Location IV
Peter	20	135	0	75
Osha	60	215	125	0
Rose	100	0	45	40
Lamark	0	90	71	0

Table 33.
Column reduction table.

Location/ Accountants	Location I	Location II	Location III	Location IV	Number of unique zero
Adidas	20	135	0	75	1
Osha	60	215	125	0	1
Rose	100	0	45	40	1
Lamark	0	90	71	0	2
Number of unique zeros	**1**	**1**	**1**	**2**	

Table 34.
Apply the straight line rule.

Accountant	Possible Branches	Assigned Branches	Efficiency Score
Peter	Location III	Location III	255
Osha	Location IV	Location IV	315
Rose	Location II	Location II	215
Lamark		Location I/Location IV Location I	190
		Total Score	975

Table 35.
Optimal assignment table.

6.2 The branch and bound technique

This method uses an iterative approach to find an optimal assignment of facilities to the available locations. The method tries to use an approach similar to the decision tree method where one branches out from the most promising or lucrative point at each stage until the final solution is reached.

Procedure

Step 1: write out the initial matrix and determine whether it is a cost or profit matrix.

Step 2: if it is a cost matrix, you need to get the least bound cost by identifying the least element in each column and adding them up.

If it is a profit matrix, get the higher bound of the profit by summing up the highest value in each column.

Step 3: Draw the branch and bound diagram and assign each personnel/task/job to the first machine/location/job and add identified least or highest element as the case may be in each selected column without repeating elements on the same row.

Step 4: repeat steps 2 & 3 until all jobs and machines have been daily assigned.

Example 1

Delight meals fast foods is planning to site four new service outlets at the four choice cities in the state, as the production and operations manager you have been given the matrix table below showing the cost of siting the service outlet in each city. Determine the most appropriate location decision for each service outlet using the branch and bound technique (see **Table 36**).

Solution.

The table shows the cost matrix so we calculate the lower bound of the total cost by adding lowest values in each column.

Hence, the lower bound of the total cost from the table above = 48 + 30 + 32 + 24 = 134.

Check and see if you can get a feasible solution from this.

Service outlet D assigned to C1 = N48.

Service outlet B assigned to C2 = N30.

Service outlet D assigned to C3 = N32.

Service outlet D assigned to C4 = N24.

From this, a feasible solution is not attained because service outlet D has been assigned to three different cities while A and C have not been assigned at all.

Therefore we have to proceed by finding a better assignment which entails branching out from our lower bound cost of 134 (see **Figure 1**).

Calculate the cost of assigning jobs to C1.

Assignment to C1

If Service outlet A is assigned to C1 == 84 + 30 + 32 + 24 = 170.

If Job B is assigned to C1 === 60 + 40 + 32 + 24 = 156.

If Job C is assigned to C1 === 60 + 30 + 32 + 24 = 146.

If Job D is assigned to C1 === 48+ 30 + 40 + 30 = 148.

Service Outlets/ Cities	C1	C2	C3	C4
A	84	70	56	42
B	60	30	40	30
C	60	50	40	30
D	48	40	32	24

Table 36.
Assignment of service outlet to C1.

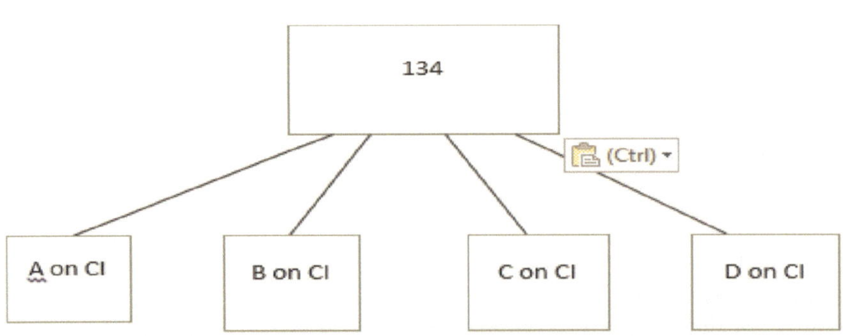

Figure 1.
Lower bound.

From the assignment above the least cost of assignment is 146 so we assign Service Outlet C to C1 (see **Figure 2**).

Allocate each of the remaining service outlets to C2 to determine the most optimal.

Assignment to C2

If Service outlet A is on C2 ⇒ (60) + 70 + 32 + 24 = 186.

If Service outlet B is on C2 ⇒ (60) + 30 + 32 + 24 = 146.

If Service outlet D is on C2 ⇒ (60) + 40 + 40 + 30 = 170.

The least cost of assignment to C2 is 146, therefore we assign Service Outlet B to City 2 (see **Figure 3**).

At this point only two cities are left C3 and C4 we can either put service outlet A on C3 and service outlet D on C4 or we can put service outlet A on C3 and A on C4. We obtain the lower bound cost for these remaining assignments as shown below. Note that service outlets C and B are already on C1 and C2 respectively.

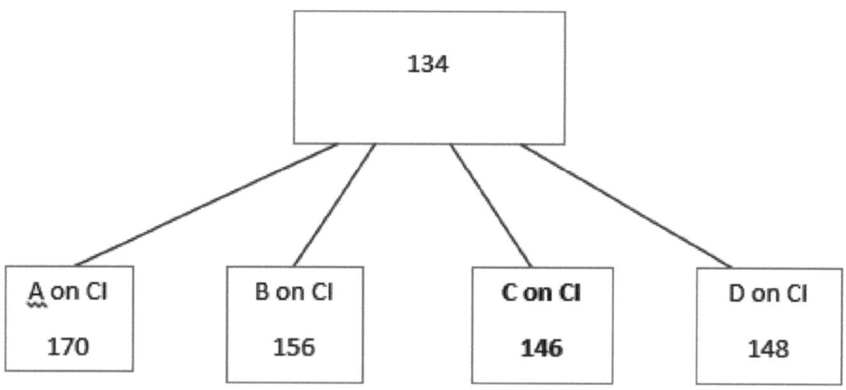

Figure 2.
Assignment of service outlet C to city 1.

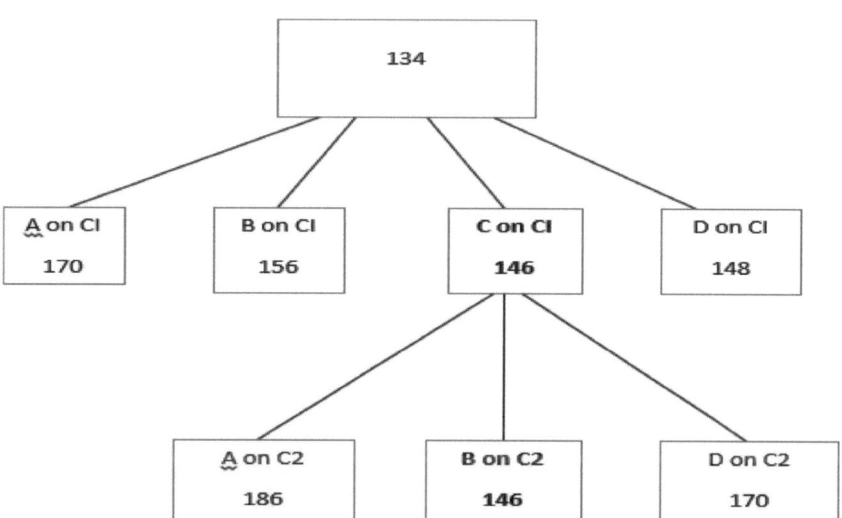

Figure 3.
Assignment of service outlet B to city 2.

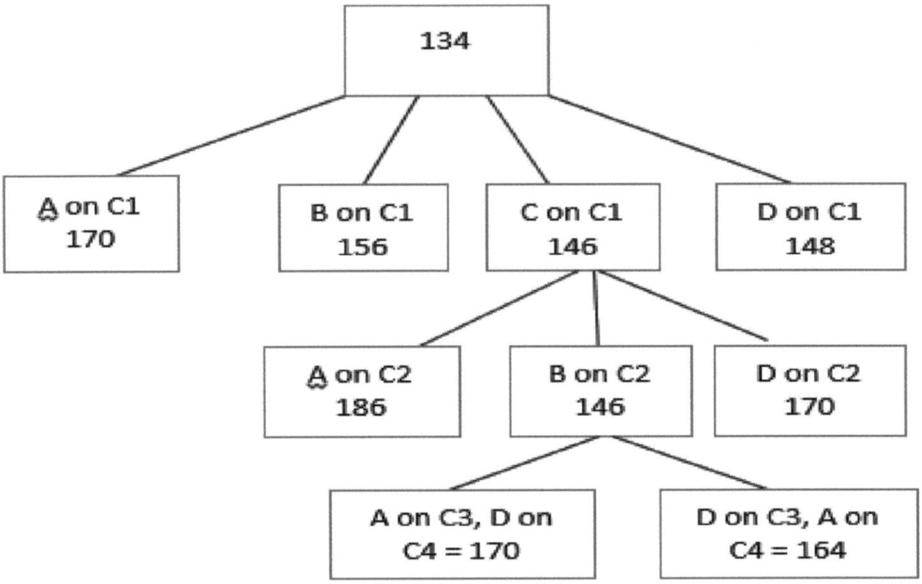

Figure 4.
Assignment of service outlets a and C.

Assignment of Service outlets to C3 & C4
Put Service outlet A on C3 and Service outlet D on C4 ⇒ (60 + 30) + 56 + 24 = 170.
Put Service outlet D on C3 and Service outlet A on C4 ⇒ (60 + 30) + 32 + 42 = 164.
Service outlet D is assigned to C3 and Service outlet A to C2.
(See **Figure 4**).

Assignment	Total
Assign Service outlet C to City 1	60,000
Assign Service outlet B to City 2	30,000
Assign Service outlet D to City 3	32,000
Assign Service outlet A to City 4	42,000
Total Cost	**164,000**

Hence the optimal assignment of Service Outlets to Cities is as shown above.
Example 2

Jobs/Machine	1	2	3	4
A	28	42	21	35
B	20	30	12	25
C	16	30	15	25
D	20	24	15	20

Table 37.
Cost matrix.

Suppose as an operations manager, the cost matrix shown above is given to you and you are required to assign the four jobs to the machines in an optional manner using the Branch and Bound method (see **Table 37**).

Solution.

Calculate for lower bound for the total cost of the assignment.

Lower bound cost = 16 + 24 + 12 + 20 = 72.

Job C is best assigned to machine 1.

Job D is best assigned to machine 2.

Job B is best assigned to machine 3.

Job D is best assigned to machine 4.

From this calculation job A is left unassigned to any machine while job D has been assigned to two machines, this is not a feasible assignment because there are multiple assignments of job D. We proceed by branching off from this lower bound cost to determine the best assignment for each machine.

Start by allocating each job in turn to machine 1 to determine the best assignment for machine 1.

Assignment on m1

(See **Figure 5**).

Put A on 1 = 28 + 24 + 12 + 20 = 84.

Put B on 1 = 20 + 24 + 12 + 20 = 79.

Put C on 1 = 16 + 24 + 12 + 20 = 72.

Put D on 1 = 20 + 30 + 12 + 25 = 87.

Therefore Job C is best assignment to machine 1.

(See **Figure 6**).

Assignment on m_2

Job A on m_2 = (16) + 42 + 12 + 20 = 90.

Job B on m_2 = (16) + 30 + 15 + 20 = 81.

Job D on m_2 = (16) + 24 + 12 + 25 = 77.

(See **Figure 7**).

Having the lowest cost of 77, job D is best assigned to m2.

Assignment on m3 and m4

Job A on 3, B on 4 = (16 + 24) + 21 + 25 = 86.

Job B on 3, A on 4 = (16 + 24) + 12 + 35 = 87.

(See **Figure 8**).

Put job A on m3 and job B on m4.

Hence, the optional assignment of the four jobs to machine that will result in the least cost assignment is as follows.

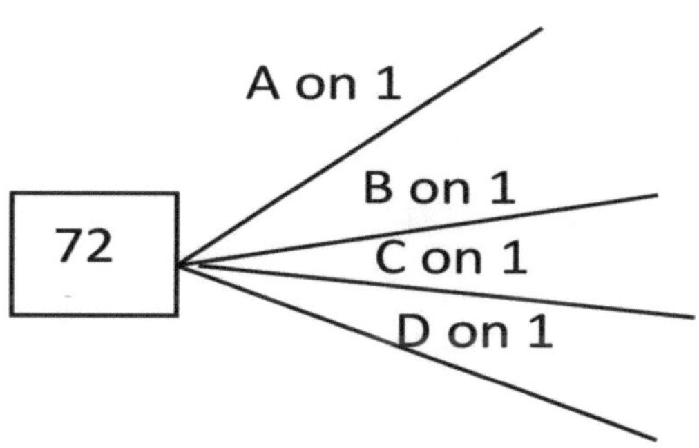

Figure 5.
Lower bound.

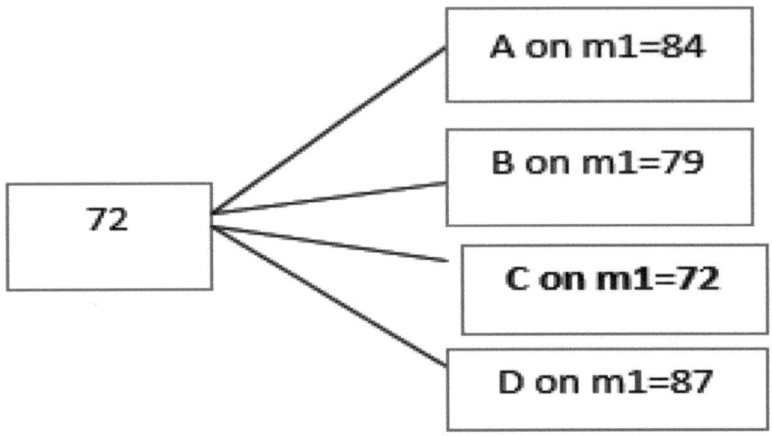

Figure 6.
Assignment of job C to machine 1.

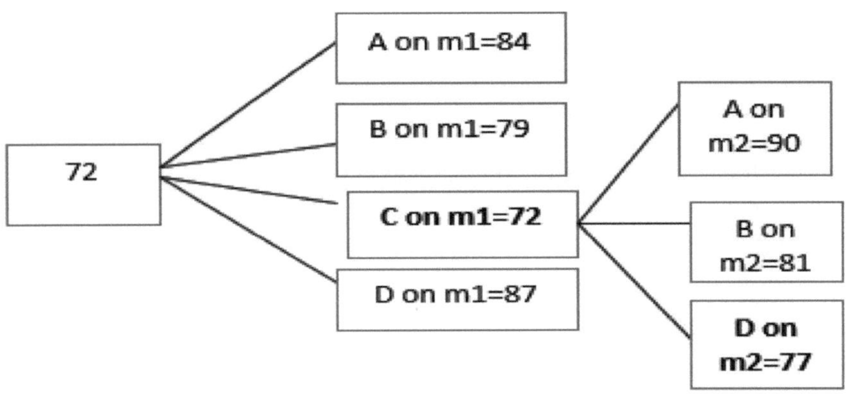

Figure 7.
Assignment of job D to machine 2.

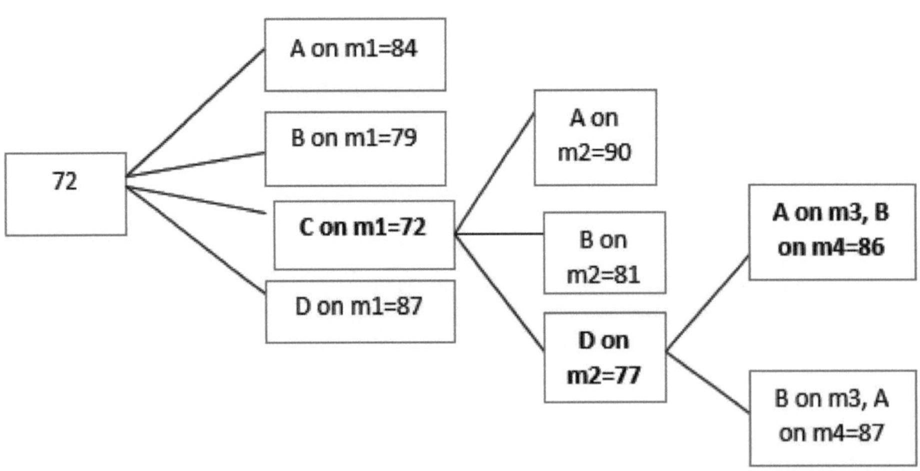

Figure 8.
Assignment of job a and B.

Assign job C to machine 1 = 16.
Assign job D to machine 2 = 24.
Assign job A to machine 3 = 21.
Assign job B to machine 4 = 25.
Total = 86.

3. Suncity a mobile phone company wishes to allocate its critical tasks to its best four operators and the cost of assigning each task to a particular operator has been provided in the table below, as the company's operations manager, advice the company on the best assignment that will maximize its total profit.

(See **Table 38**).

Task/Operators	1	2	3	4
A	130	190	280	200
B	280	310	300	150
C	170	200	210	400
D	120	260	280	210

Table 38.
Profit matrix.

Solution.

In a maximization situation the higher bound of the given matrix is calculated by adding the highest value in each column, from this value we can then branch off to determine the most feasible assignment. It can be observed that all the subsequent profit values after the higher bound is obtained are either equal to, or lower than the higher bound.

Higher bound = 280 + 310 + 300 + 400 = 1290.
The initial assignment will be.
Assign task B to Operator 1.
Assign task B to Operator 2.
Assign task B to Operator 3.
Assign task C to Operator 4.

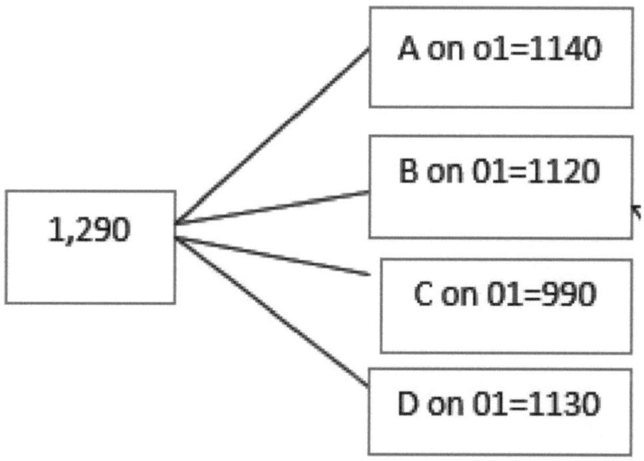

Figure 9.
Lower bound.

This is not a feasible assignment because we have task B assigned to operator 1, 2 and 3, task C to operator 4, while task A and D have not been assigned to any operator. Hence we branch off from the highest bound to determine the optimal assignment to each operator (see **Figure 9**).

Assignment of task to Operator 1

Assign task A to operator 1 = = 130 + 310 + 300 + 400 = 1140.

Assign task B to operator 1 = = 280 + 260 + 280 + 400 = 1220.

Assign task C to operator 1 = = 170 + 310 + 300 + 210 = 990.

Assign task D to operator 1 = = 120 + 310 + 300 + 400 = 1130.

Therefore, we will assign task B to operator 1 because it yields the highest profit of 1220 (see **Figure 10**).

Assignment of task to operator 2

Assign task A to operator 2 = = (280) + 190 + 280 + 400 = 1150.

Assign task C to operator 2 = = (280) + 200 + 280 + 210 = 970.

Assign task D to operator 2 = = (280) + 260 + 280 + 400 = 1220.

Therefore assign task D to operator 2 (see **Figure 11**).

Assignment of task to operator 3 and 4

Assign task A to operator 3 and task C to operator 4 = (280 + 260) + 280 + 400 = 1220.

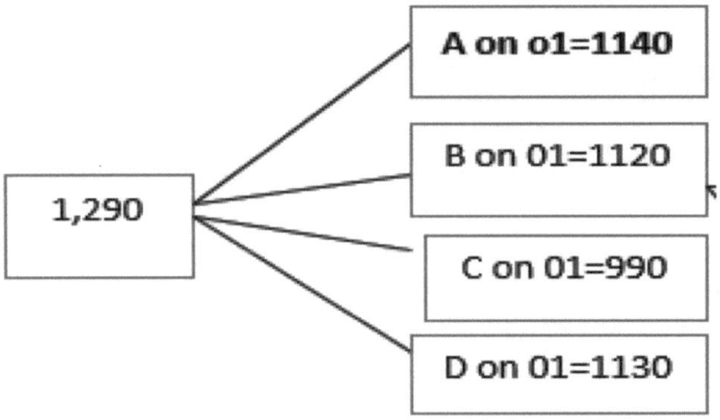

Figure 10.
Assignment of task B to operator 1.

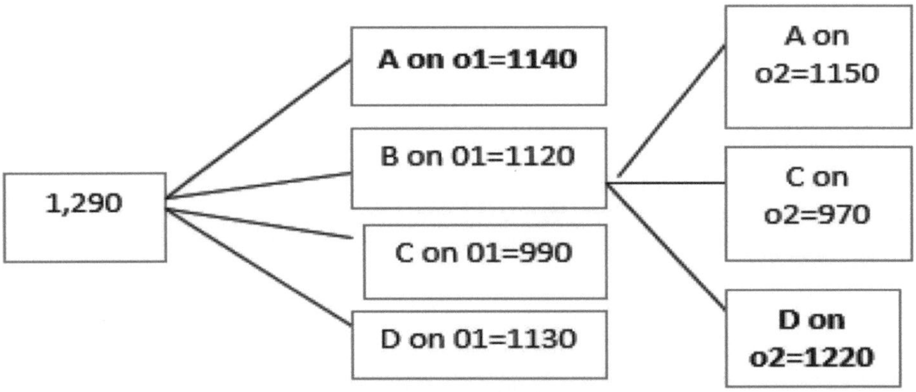

Figure 11.
Assignment of task D to operator 2.

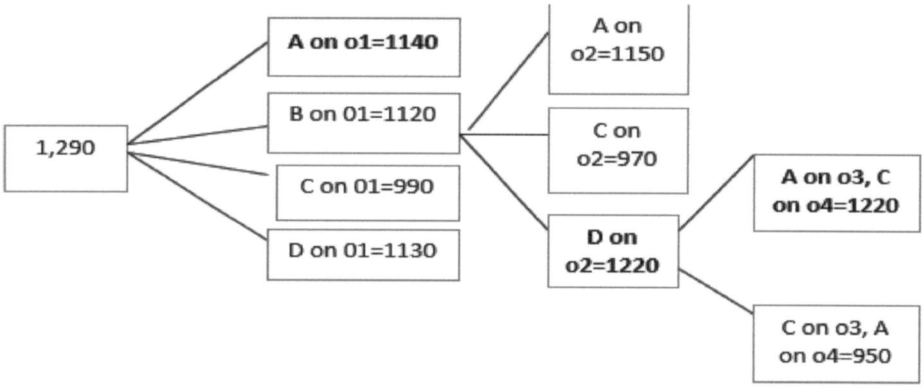

Figure 12.
Assignment of task a and C to operator 3 and 4 respectively.

Optimal Assignment of Task to Operator	Profit
Assign task B to operator 1	280
Assign task D to operator 2	260
Assign task A to operator 3	280
Assign task C to operator 4	400
Total Profit	**1220**

Table 39.
Assignment table.

Assign task C to operator 3 and task A to operator 4 = (280 + 260) + 210 + 200 = 950.

Hence we assign task A to operator 3 and task C to operator 4 (see **Figure 12** and **Table 39**).

7. Conclusion

Facility location is an important aspect of production and operations management. Where to locate a new plant or facility is an expensive decision that is not frequently made. Therefore, caution must be made not to site facilities in non-attractive or less optimal locations, as this will affect the efficiency levels of the production and distribution system of any organization, and in-turn its survival. In determining an optimal facility location, it is observed that the least cost or highest profit may not always be feasible, this is because in real life situations, there are some limitations to choosing the most appropriate allocation. Due to such challenges, there may be need for trade-offs. Hence, the optimal assignment represents the most feasible situation where all the facilities have been suitably assigned locations. The Hungarian and branch and bound method are most suitable approaches to solving location problems.

Practice Questions

1. Given the cost matrix show below, you are required to assign the four Engineers to the four sites in an optimal manner using branch and bound method of liner assignment (see **Table 40**).

Machine/ Jobs	1	2	3	4
A	28	41	18	15
B	42	31	32	38
C	7	41	40	50
D	58	35	19	16

Table 40.
Cost matrix.

Machine/ Jobs	1	2	3	4
A	100	130	150	180
B	140	140	180	100
C	170	80	70	160
D	180	160	110	90

Table 41.
Profit matrix.

2. Given the profit matrix below assign the operators to the machines in such a way that will maximize the total profit (see **Table 41**).

Nomenclature list

M	denotes rows
N	denotes columns
//	number of straight lines drawn on a matrix
=	equal to
≠	not equal to

Author details

Nneoma Benita Amos* and Edafe Bawa Dogo
Department of Business Administration and Marketing, Babcock University, Ilishan-Remo, Ougn State, Nigeria

*Address all correspondence to: amosn@babcock.edu.ng

References

[1] Adeleke OJ, Olukanmi DO. Facility location problems: Models, techniques, and applications in waste management. Recycling. 2020;**5**(10):2-20

[2] Alenezy EJ. Solving capacitated facility location problem using Lagrangian decomposition and volume algorithm. Advances in Operations Research. 2020;**2020**:1-7. DOI: 10.1155/2020/5239176

[3] Banjoko SA. Production and Operations Management. Lagos: Wisdom Publishers ltd.; 2012

[4] Gupter S, Starr M. Production and Operations Management Systems. New York: Taylor & Francis Group; 2014

[5] Panneerselvam R. Production and Operations Management. 3rd ed. PHI learning private limited: New Delhi; 2012

[6] Russel RS, Taylor BW. Operations Management: Creating Value along the Supply Chain. 7th ed. John Wiley and Sons, Inc.: United States of America; 2011

[7] Stevenson WJ. Operations Management. 12th ed. McGraw Hill: New York; 2015

Chapter 4

Forecasting Techniques and Accuracy of Sales Performance of a Grocery Store

John Kolade Obamiro, Wasiu Oluwasegun Kehinde, Jonathan Ehimen Ekpudu and Oluwasegun David

Abstract

This study examined forecasting techniques and accuracy of sales performance of a grocery store. The survey and ex-post facto research designs were employed. The study employed Naive, exponential smoothing, least square, moving average, multiplicative decomposition and additive decomposition methods to forecast sales using sales data generated from January, 2017 to August, 2018 while the forecast accuracy was evaluated. The outputs of the methods were compared to determine the optimal forecasts and the demand for the next 14 months using the multiplicative and additive methods. The MAPE for Naïve, Exponential smoothing, Least square, Moving average, Multiplicative and Additive decomposition are 10.9, 10.2, 8.6, 11.9, 7.64, and 7.66%. By comparison, it was observed that multiplicative decomposition is the best technique because it generates the optimal forecast accuracy of 7.64%. Thus, the study recommends that the grocery outlet should maintain effective and reliable sales records for accurate forecasts.

Keywords: forecasting techniques, forecast accuracy, sales performance and grocery

1. Introduction

One of the major functions of management is to make decisions that will bring about a brighter future for the organization. The future can be predicted using past data or the formation of opinion [1]. Chao et al., [2] assert that an important role in the business planning process has led to extensive research as far as forecasting is concerned. Forecasting is defined as the art and science of predicting future events. It may involve the use of hard data (historical or past data) and/or soft data (experience, intuition, personal value, and judgment) in projecting into the future with some sort of mathematical model [3]. Ayers and Odegaard [4] posit forecasts play a crucial role in business management and strategic planning. The management decisions taken at every level of a business are either directly or indirectly related to the forecasts it has made. Without useful predictions, planning and control activities cannot be accomplished effectively. Forecasting is of great significance for organizations including grocery/retail companies because all firms will want to minimize the capital tied in their stocks [5], and meet customers' demands, hence, profitability.

In the present turbulent market environment, forecasting the right volume of demand and the future development has an irreplaceable role in the management of every retail business [6]. The objective of groceries/retailers worldwide is to provide the right products at the right time and maintaining a healthy stock [7]. Retailers considering investing in advanced solutions should get acquainted with forecast accuracy [8]. Kumar and Dalgobind [9], viewed forecasting as an important part of retail decision-making. Forecasting helps to improve the customer experience through shorter delivery times, and it also allows the company to find better ways to use inventory-bound funds [10]. Paták et al. [11] posit that accurate demand forecasting is a prerequisite for efficient and effective customer satisfaction. It makes it possible to efficiently use resources and reduce costs associated with excess, or lack of goods within the supply chain. Accurate forecasting of future retail sales can contribute to gains in the efficiency and efficacy of operations carried out in retail businesses and supply chains [4].

Groceries have been confronting high customers' demands, overstock, and stock out of inventory for years. This is getting worst with newer non-traditional channels like phoning and online purchase and delivery apps fueling the growth of grocery retailers, traditional stores must adapt or be left behind [12]. Due to the hyper-competitive industry of grocery, customer loyalty is dying, so experiencing stock-outs might lead to losing customers. Corsten and Gruen, [13] found that customers whom we once believed to either purchase substitutes or delay purchases when their preferred products are out of stock are more likely to switch stores and never come back. Stock-outs cause a direct loss of potential sales and lead to dissatisfied customers. The stock-out of individual items not only has a negative impact on its sales but also on the sales of the whole product category. In preventing stock-outs, retailers often have large safety stocks [14] which substantially increases costs and reduces profit. In some cases, despite the safety stocks, the chance of a stock-out is huge, with potentially a big impact on customer satisfaction and sales [15, 16]. The best preparation for the dilemma is to emphasize accurate demand forecast so they can carry the right supply for each product. Accurate forecast helps the Grocers to resolve the cost and service trade-off by balancing the loss due to stock-outs and the cost of safety stocks. Despite the importance of accurate demand forecasts compare to demand forecasts which are mostly wrong and misleading, forecast accuracy in the retail sector has not been adequately covered in the literature and by practitioners which is the basis behind this study. Also, the study used real and accurate sales data (sales history) against randomly generated demand data [17]. The more real and valid the data to be used for forecast, the more accurate is the outcome (demand forecast) [18].

Another issue confronting the practitioners or management is which forecasting techniques can generate reliable sale forecasts. Most grocery retailers in Nigeria hardly do forecast and the few ones use qualitative techniques; where experiences, opinions, and judgments of sales are largely used to determine a sales forecast. Even when quantitative forecasting was employed, the Naïve forecasting method is the most commonly used by grocery retailers. The reasons are that most managers are not conversant with quantitative forecasting techniques or do not want to put effort or do not see the benefits of quantitative techniques; using historical sales to determine sales forecasts. Moon et al. [19] cited by Orrebrant and Hill [20] mentions that it can be both costly and time-consuming to rely only on quantitative techniques when forecasting. He further advised that companies should therefore evaluate how to adopt different forecasting techniques to increase the sales forecast accuracy. Based on these premises and given the essential role of grocery retailing and the relevance of accurate demand forecast, this study adopted quantitative forecasting methods such as Naive, Exponential Smoothing, Least Square, Moving Average, Multiplicative, Decomposition, and Additive Decomposition Methods to determine demand forecast from actual sales history for a given period. Also, the demand forecast accuracy was determined using Mean Absolute Deviation (MAD), Mean Squared Error (MSE), and Mean Absolute Percentage Error (MAPE).

2. Literature review

2.1 Conceptual framework

Forecasting is a tool that helps decision-makers to cope with the uncertainties of the future through prediction. Forecasting techniques can be divided into qualitative and quantitative. The qualitative forecasting technique uses soft information such as personal experience, value, intuition, and judgment to predict future sales. This technique is assumed to be subjective and biased owing to a lack of scientific proof. Some commonly used qualitative forecasting methods are; Delphi, jury of executives, consumers' survey, and sales force opinions. Kumar and Dalgobind [9] gave the account of the circumstances where qualitative forecasting methods are preferable to quantitative ones as when data are not available or insufficient or are known to be unreliable, when it is not possible to construct a suitable numerical model and when the time is insufficient to initiate and operate a quantitative analysis.

The quantitative forecasting model uses hard data (historical data) to estimate future demands with some sort of mathematical model [21]. Quantitative forecasting method is appropriate when past data are available, sufficient, reliable and when it is possible to use mathematical models. This forecasting method involves trend projection using naïve method, sales trend analysis, causal, time series, moving average, exponential smoothing, and least square method. Generally, the forecast is usually classified by the time duration and uses. Short-term forecast which is the responsibility of first-level managers covers a period of up to one year involving daily operating activities. A medium-term forecast addresses cash budgeting, production and aggregate demand planning aggregate which is carried out by middle level manager for a period of one to three years. Long-run forecast involves taking strategic decisions that affect organizational strategic objectives, strategies, and policies. This is mostly the responsibility of top management.

2.2 Empirical studies

Some studies addressed the grocery retailing and maintaining healthy inventory through accurate demand forecasting involving different products in several countries [4, 10–12, 14, 20] using different simple and advanced forecasting methods (either qualitative, quantitative and hybrid) to model sales data in order to predict the future demand which is essential for the planning process. Aras et al. [22] used state space models, ARIMA and ARFIMA models, neural networks, and Adaptive Network-based Fuzzy Inference System (ANFIS) to determine sales forecasts for a global furniture retailer operating. The findings revealed that the single model in isolation did not outperform all the others across all the time series investigated. However, most of the combined forecasts examined could achieve statistically significant increases in forecasting accuracy compared with individual models and with the forecasts generated by the company's current system. Chu and Peter [23] compared linear and nonlinear models for retail sales forecasting and established that nonlinear models are able to outperform their linear counterparts. Sapirstein [24] adopted Brown's exponential smoothing method to improve the sale of pharmaceutical products, taking into account the fact that Romania is and will continue to be a market of high potential in selling pharmaceutical goods. The study advised that the managers of health care institutions should use the new information and communication technologies in order to plan the activities that will contribute to reaching their proposed objectives, based on accurate forecasting patterns. Lee et al., [17] concluded that the overall best model in their study was the neural network built on deseasonalized time series data. Sahu and Kumar [25] study on forecasting models and their applications for sales forecasting of sterilized flavored milk in Chhattisgarh. The study adopted naïve model, moving average, double moving average, simple exponential smoothing; and semi average method to analyze weekly data spreading over October 2011 to October 2012, on the sales of sterilized milk, and the accuracy of the forecasting method was measured using mean Forecast Error (MFE), Mean Absolute Deviation (MAD), Mean Square Error (MSE), root mean square error (RMSE).

3. Materials and methods

This study used a survey and ex-post facto research design. The Secondary data of sales history from the Grocery Bazaar Akesan branch for 20 months was used. Various quantitative forecasting techniques such as Naive, Exponential Smoothing, Least Square, Moving Average, Multiplicative Decomposition, and Additive Decomposition Method were employed to analyze twenty months of data (from January 2017 to August 2018) collected to predict the demand forecasts for 14 months from September 2018 to October 2019. The demand forecasts were subjected to accuracy tests to identify errors made. This study compares the Mean Absolute Deviation (MAD), the mean absolute percentage error (MAPE), and the mean square error (MSE) to evaluate the performance of the forecasting methods and identify the least error among the methods. The data generated were analyzed using operations research software (POM QM).

3.1 Data presentation

Data used in this study consist of actual sales and product details from the chosen company (Grocery, Bazaar). The company has more than 80 brands. Sales data for twenty months (20) from 1st January, 2017 to 31st August 2018 were collected. Raw sales data of different brands (such as bakery, baking supplies, beverages, cosmetics, cleaning supplies, dairy, frozen food, meat and fish, ready-made

S/N	Year	Quarter	Months (x)	Actual Sales (y) (N)
1	2017	1st	January	90,513,640
2			February	85,809,019
3			March	91,651,973
4		2nd	April	104,616,584
5			May	92,348,534
6			June	89,874,755
7		3rd	July	86,988,783
8			August	82,364,715
9			September	82,067,415
10		4th	October	74,056,416
11			November	71,515,133
12			December	119,649,394
13	2018	1st	January	78,872,065
14			February	70,656,413
15			March	84,965,677
16		2nd	April	84,404,820
17			May	82,307,423
18			June	80,003,438
19		3rd	July	75,715,781
20			August	85,783,239

Source: Company's sales, January 2017–August 2018.

Table 1.
20 months sales' figure for the retail outlet.

meals, sweets, snacks, and so on), were computed (see **Table 1**) and subjected to various forecasting techniques.

4. Data analysis

The six (6) forecasting methods (Naive, exponential smoothing, least square, moving average, multiplicative decomposition, and additive decomposition method) were used to determine sales forecasts from September 2018 to October 2019. The performance of the forecasting methods was evaluated with forecast accuracy measures (MAD, MAPE, and MSE). This study dwelled more on forecasting using six techniques and 20 Months sales history. This research is an extension of studies conducted by Sahu and Kumar [25] and Obamiro [26]. Obamiro's [26] study predicted and evaluated demand forecast accuracy for twelve months using three quantitative forecasting methods (see **Table 2**).

Figures 1–3 depict the results obtained from the use of naive, exponential smoothing, and moving average methods of the forecast models. The MAPE for the three methods are 10.93%, 10.7%, and 11.91% respectively and this makes exponential smoothing the best method with minimum sales of N82, 171,270 (see **Tables 3** and **4**).

	Naive Method	Exponential Smoothing	Least Square	Moving Average
Measure	**Value**			
Error Measures				
Bias (Mean Error)	−248968.4	−878145.3	−10.8	−412484.4
MAD (Mean Absolute Deviation)	9,861,447	9,042,428	7586159.0	10,519,060
MSE (Mean Squared Error)	255,953,800,000,000	173,845,100,000,000	120,610,000,000,000	227,368,300,000,000
Standard Error (denom = n^{-2} = 17)	16,913,490	13,939,060	11,576,320	15,993,410
MAPE (Mean Absolute Percent Error)	10.939%	10.17%	8.633%	11.914%
Forecast				
next period	85,783,240	82,171,270	85,708,270	80,749,510

Source: Field work, 2018.

Table 2.
Result for naive method, exponential smoothing, Least Square, and moving average.

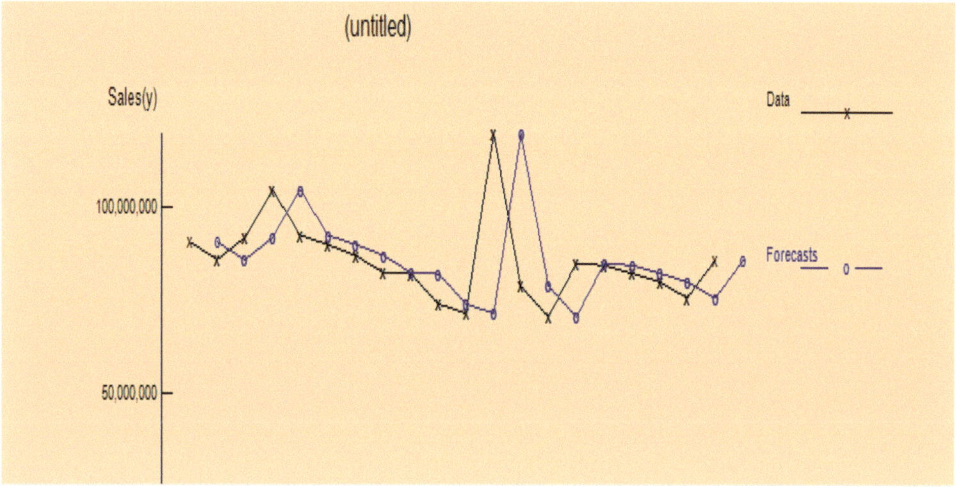

Figure 1.
Graph of naive method. Source: Field work, 2018.

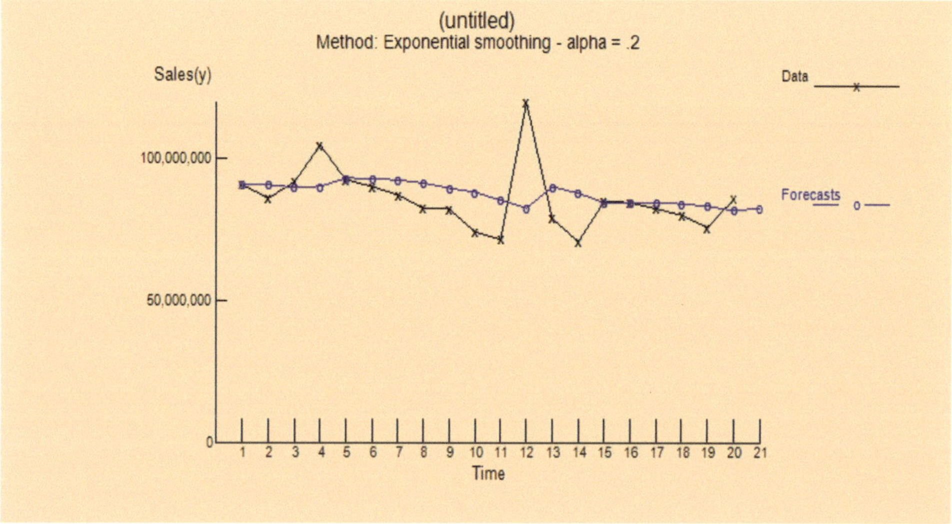

Figure 2.
Graph of exponential smoothing. Source: Field work, 2018.

5. Discussions

The analysis reveals the MAD, MSE, and MAPE of the six methods employed. MAPE for Naïve, Exponential Smoothing, Least Square, Moving Average, Multiplicative and Additive Decomposition is 10.9%, 10.2%, 8.6%, 11.9%, 7.64%, and 7.66% respectively. By comparison, it was observed that multiplicative decomposition seems to be preferable among the six techniques. This is because it generates the optimal forecast accuracy of 7.64%. This is contrary to the works of Sahu and Kumar [25] and Obamiro, [26] which recommended exponential smoothing as the best forecasting technique. Forecasts for the next period which is September 2018 sales, Naïve shows 85,783,240, Exponential Smoothing gave 82,171,270, Least Square has

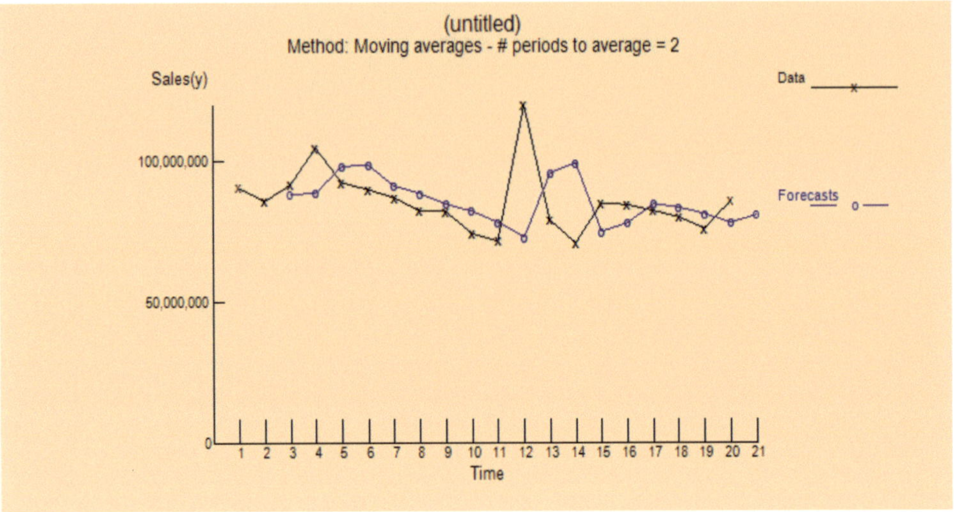

Figure 3.
Graph of moving average method. Source: Field work, 2018.

Measure	Value	Future Period	Unadjusted Forecast	Seasonal Factor	Adjusted Forecast
Error Measures		21	78,798,730	.971	76,533,570
Bias (Mean Error)	7451.2	22	78,099,960	1.019	79,571,400
MAD (Mean Absolute Deviation)	6,710,237	23	77,401,200	.971	75,176,210
MSE (Mean Squared Error)	100,779,100,000,000	24	76,702,430	1.019	78,147,540
Standard Error (denom = n-2-2 = 16)	11,223,810	25	76,003,670	.971	73,818,860
MAPE (Mean Absolute Percent Error)	7.641%	26	75,304,910	1.019	76,723,690
Regression line (unadjusted forecast)		27	74,606,140	.971	72,461,500
Sales(y) = 93,472,760		28	73,907,380	1.019	75,299,820
−698763.6 * time		29	73,208,620	.971	71,104,140
Statistics		30	72,509,860	1.019	73,875,980
Correlation coefficient	.406	31	71,811,090	.971	69,746,790
Coefficient of determination (r^2)	.165	32	71,112,330	1.019	72,452,110
		33	70,413,560	.971	68,389,440
		34	69,714,800	1.019	71,028,260

Source: Field work, 2018.

Table 3.
Result for multiplicative composition method.

Measure	Value	Future Period	Unadjusted Forecast	Seasonal Factor	Adjusted Forecast
Error Measures		21	78,608,890	−2499435.0	76,109,460
Bias (Mean Error)	1.2	22	77,913,220	2089101.0	80,002,320
MAD (Mean Absolute Deviation)	6,725,763	23	77,217,550	−2499435.0	74,718,120
MSE (Mean Squared Error)	100,542,500,000,000	24	76,521,880	2089101.0	78,610,980
Standard Error (denom = n-2 = 18)	10,569,480	25	75,826,210	−2499435.0	73,326,780
MAPE (Mean Absolute Percent Error)	7.663%	26	75,130,540	2089101.0	77,219,640
Regression line (unadjusted forecast)		27	74,434,860	−2499435.0	71,935,430
Sales(y) = 93,217,960		28	73,739,200	2089101.0	75,828,300
−695670.1 * time		29	73,043,530	−2499435.0	70,544,100
Statistics		30	72,347,860	2089101.0	74,436,960
Correlation coefficient	.408	31	71,652,190	−2499435.0	69,152,760
Coefficient of determination (r^2)	.166	32	70,956,520	2089101.0	73,045,620
		33	70,260,850	−2499435.0	67,761,420
		34	69,565,180	2089101.0	71,654,280

Source: Field Work, 2018.

Table 4.
Result for additive composition method.

85,708,270, Moving Average shows 80,749,510, and Multiplicative has 76,533,570, while Additive Decomposition is 76,109,460. This shows that additive decomposition is optimal in terms of forecast accuracy.

6. Conclusion

Kumar and Dalgobind [9] gave situations where qualitative methods are preferable to quantitative ones and they are when data are insufficient or are known to be unreliable, when it is not possible to construct a suitable numerical model and when the time is insufficient to initiate and operate a quantitative analysis. In groceries, comprehensive historical data (past sales) can be obtained by proper documentation of daily sales and this makes quantitative techniques more appropriate. Data used for this study were obtained from the grocery on the assumption that the sale figures were properly recorded. The performance of the forecasting methods (Naïve, exponential smoothing, least square, moving average, multiplicative, and additive decomposition method) in terms of accuracy of sales forecasts using MAD, MSE, and MAPE were determined. The findings reveal that the multiplicative method has the least forecast error and is hence more accurate. One of the limitations of this study is that the analysis was limited to chosen forecasting methods ignoring the complex and sophistication methods. This limitation has no effect on the reliability and validity of the demand forecast and its accuracy. Therefore, it is recommended that retail outlets should maintain regular sales and inventory records. This becomes easier if the operating system can be computerized but it could be expensive to operate. Also, the operators should determine demand forecast scientifically with multiplicative decomposition method (quantitative) forecasting techniques with past data instead of guessing estimates from using qualitative forecasting techniques which rely on soft information such as personal experience, intuition, values, and opinions.

Author details

John Kolade Obamiro[1*], Wasiu Oluwasegun Kehinde[1], Jonathan Ehimen Ekpudu[2] and Oluwasegun David[3]

1 Faculty of Management Sciences, Department of Business Administration, Lagos State University, Ojo, Nigeria

2 Department of Business Administration, Federal University of Agriculture Abeokuta, College of Management Sciences, Ogun State, Nigeria

3 Department of Business Administration, Bowen University, College of Management Sciences, Iwo, Osun State, Nigeria

*Address all correspondence to: johnobamiro@gmail.com

References

[1] Chindia EW, Wainaina G, Kibera FN, NF and GP Pokhariyal, PG Forecasting techniques, operating environment and accuracy of performance forecasting for large manufacturing firms in Kenya. International Journal of Managerial Studies and Research. 2014;**2**(7):83-100

[2] Chao Y, Jimei Y, Brian P, Scott C, Jia D. Forecasting human dynamics from static images. In: Proceedings of the IEEE Conference on Computer Vision and Pattern Recognition. 2017. pp. 548-556

[3] Türkşen O, Tez M. An application of nelder-mead heuristic-based hybrid algorithms: Estimation of compartment model parameters. International Journal of Artificial Intelligence. 2016;**14**(1):112-129

[4] Ayers JB, Odegaard MA. Retail Supply Chain Management. 2nd ed. Boca Raton: CRC Press; 2017

[5] Levy M, Weitz B. Retailing management. McGraw-Hill Education; 2011

[6] Jain A, Menon MN, Chandra S. Sales forecasting for retail chains. San Diego, CA: UC San Diego Jacobs School of Engineering; 2015

[7] Gemmink M. A model to prevent stockouts in retail using time series sales forecasting. In: A Paper Presented at 27th Twente Student Conference on IT, Faculty of Electrical Engineering, Mathematics and Computer Science. University of Twente; 2017

[8] Belt T. When is forecast accuracy important in the retail industry? Effect of key product parameters. In: Master's Thesis Submitted to the Department of Industrial Engineering and Management. Aalto University School of Sciences; 2017

[9] Kumar R, Dalgobind M. Application of proper forecasting technique in juice production: A case study. Global Journal of Researches in Engineering. 2014;**13**(4):42-47

[10] Järvinen S. Improving 3-month sales forecasting for the sales unit. In: Being a master's Thesis Submitted to the Department of Industrial Management. Helsinki Metropolia University of Applied Sciences; 2017

[11] Paták M, Branska L, Pecinova Z. Demand forecasting in retail grocery stores in the Czech Republic. In: 2nd International Multidisciplinary Scientific Conference on Social Sciences and Arts SGEM. 2015. pp. 693-700

[12] Kini D, Mathur U, Viswanathan M. Dynamic race prediction in linear time. ACM SIGPLAN Notices. 2017;**52**(6): 157-170

[13] Corsten D, Gruen T. Stock-outs cause walkouts. Harvard Business Review. 2001;**82**(5):26-28

[14] Xia M, Zhang Y, Weng L, Ye X. Fashion retailing forecasting based on extreme learning machine with adaptive metrics of inputs. Knowledge-Based Systems. 2012;**36**:253-259

[15] Ehrenthal JC, Stolzle W. An examination of the causes for retail stockouts. International Journal of Physical Distribution & Logistics Management. 2013;**43**(1):54-69

[16] Helm R, Hegenbart T, Endres H. Explaining costumer reactions to real stockouts. Review of Managerial Science. 2013;**7**(3):223-246

[17] Lee HS, Kim H, Park H, Kang P. Pre-launch new product demand

forecasting using the bass model: A statistical and machine learning-based approach. Technological Forecasting and Social Change. 2014;**86**:49-64

[18] Lawrence M, O'Connor M, Edmundson B. A field study of sales forecasting accuracy and processes. European Journal of Operational Research. 2000;**122**(1):151-160

[19] Moon S, Simpson A, Hicks C. The development of a classification model for predicting the performance of forecasting methods for naval spare parts demand. International Journal of Production Economics. 2013;**143**(2): 449-454

[20] Orrebrant R, Hill A. Increasing sales forecast accuracy with technique adoption in the forecasting process. 2014

[21] Özlem AKAY. A computational method based on interval length for fuzzy time series forecasting. NATURENGS. 2016;**2**(1):22-33

[22] Aras S, Deveci Kocakoç I, Polat C. Comparative study on retail sales forecasting between single and combination methods. Journal of Business Economics and Management. 2017;**18**(5): 803-832

[23] Chu CW, Peter G. A comparative study of linear and nonlinear models for aggregate retail sales forecasting. International Journal of Production Economics. 2003;**86**:217-231

[24] Sapirstein P. Accurate measurement with photogrammetry at large sites. Journal of Archaeological Science. 2016;**66**:137-145

[25] Sahu PK, Kumar R. Demand forecasting for sales of milk product (Paneer) in Chhattisgarh. International Journal of Inventive Engineering and Sciences. 2013;**1**(9):10-13

[26] Obamiro JK. Demand Forecasting and Measuring Forecast Accuracy in a Pharmacy. Unpublished manuscript. 2018

Chapter 5

Performance Measurement Using Deterministic and Stochastic Multiplicative Directional Distance Functions

Yu Zhao

Abstract

Performance measurement is essential for fostering continuous improvement of the production and operation management in a firm or organization. We consider a deterministic scenario based on a flexible structure of production technology and establish a multiplicative relationship between the generalized multiplicative directional distance function (GMDDF) and geometric distance function (GDF). We also introduce a stochastic multiplicative directional distance function (SMDDF). Based on a stochastic scenario, the SMDDF can be estimated by the method of convex nonparametric least squares. As an illustrative application, we investigate the productive performance of Japanese life insurance companies using a panel dataset spanning 2016 to 2020.

Keywords: performance measurement, stochastic multiplicative directional distance function, data envelopment analysis, convex nonparametric least squares, Japanese life insurance companies

1. Introduction

The purpose of this study is to establish the relationship between the generalized multiplicative directional distance function (GMDDF, [1]) and the geometric distance function (GDF, [2]), and to develop a stochastic multiplicative directional distance function (SMDDF) for performance measurement. The GMDDF and GDF are commonly measured by data envelopment analysis (DEA, [3]), a deterministic nonparametric method applicable to multiple-input and multiple-output production technologies. Although the GMDDF and GDF have a multiplicative nature, the original GDF is measured based on a convex production possibility set, whereas the GMDDF is developed with a non-convex production possibility set. From a practical perspective, the convexity assumption may be strong if significant degrees of increasing returns exist. To address a flexible structure of production technology, previous studies in DEA also studied a non-convex production possibility set based on a geometric convexity assumption [1, 4]. The methodology of this study is based on a geometric convex production possibility set, which implicitly models a piecewise log-linear technology. Moreover, it is well known that a one-to-one correspondence relationship exists between the piecewise log-linear technology and the conventional piecewise linear technology. This makes it possible to apply the conventional DEA methods to approach the multiplicative models developed for a geometric convex production possibility set. Therefore, measuring the multiplicative type distance functions is of practical and theoretical interest.

It is also well known that DEA commonly discounts stochastic noise, suggesting that any deviations from the frontier (e.g., gauging the distance to the boundary of the production technology) can be considered a measure of pure inefficiency. By contrast, stochastic frontier analysis (SFA, [5]), a general stochastic parametric approach, accounts for stochastic noise by treating all deviations from the frontier as aggregations of both inefficiency and noise. However, compared with the flexibility of nonparametric measurements, because SFA is a parametric methodology, it relies heavily on an accurately prespecified functional form for production technology. Recently, stochastic nonparametric envelopment of data (StoNED) has been introduced in the literature on efficiency analysis. The StoNED approach relies on the method of convex nonparametric least squares (CNLS), which is a nonparametric regression technique [6]. Both SFA and StoNED are based on a convex production possibility set. This study introduces a stochastic multiplicative directional distance function (SMDDF) based on the geometric convex production possibility set and further proposes an estimation model using CNLS.

The methodology of this study consists of two basic scenarios: deterministic and stochastic. These are performed simultaneously to analyze the productive performance of Japanese life insurance companies. The application demonstrates the advantage of using a combination of GMDDF/GDF and SMDDF, enabling comprehensive understanding of the Japanese life insurance industry.

2. Preliminaries

Consider a sample of n decision-making unitis (DMUs). Denote $\boldsymbol{x}_j = (x_{1j}, \dots, x_{mj})^T \in \mathbb{R}^m_{++}$ and $\boldsymbol{y}_j = \left(y_{1j}, \dots, y_{sj}\right)^T \in \mathbb{R}^s_{++}$ as positive input and output vectors of DMU$_j$ ($j = 1, \dots, n$), respectively. The superscript T stands for a vector transpose. A production technology that transforms inputs into outputs can be defined by the following production possibility set:

$$P := \left\{ (\boldsymbol{x}, \boldsymbol{y}) \in \mathbb{R}^m_{++} \times \mathbb{R}^s_{++} | \boldsymbol{x} \text{ can produce} \boldsymbol{y} \right\},$$

where P contains the input and output vectors $\left(\boldsymbol{x}_j, \boldsymbol{y}_j\right)$ for all $j = 1, \dots, n$. The conventional DEA-based production technology assumes that P is a closed, bounded, and convex set that satisfies free disposability (see, e.g., [7, 8]). In particular, the ordinary convexity assumption, which is described as below,

[A1] If $(\boldsymbol{x}_1, \boldsymbol{y}_1) \in P$ and $(\boldsymbol{x}_2, \boldsymbol{y}_2) \in P$, then $(\lambda \boldsymbol{x}_1 + (1-\lambda)\boldsymbol{x}_2, \lambda \boldsymbol{y}_1 + (1-\lambda)\boldsymbol{y}_2) \in P$ for all $\lambda \in [0, 1]$

requires the marginal products to be nonincreasing. However, this cannot be used to understand productive performances characterized by significant degrees of increasing returns so that the marginal products of inputs may be low with a small-scale economy and may increase sharply as the scale expands beyond a threshold (see, e.g., [9]). To allow a flexible structure of production technology, previous studies [1, 4] postulated a geometric convexity assumption, as below:

$$[A1'] \text{ If } (\boldsymbol{x}_1, \boldsymbol{y}_1) \in P \text{ and } (\boldsymbol{x}_2, \boldsymbol{y}_2) \in P, \text{ then } (\boldsymbol{x}_1, \boldsymbol{y}_1)^{\lambda} \otimes (\boldsymbol{x}_2, \boldsymbol{y}_2)^{1-\lambda} \in P \text{ for all } \lambda \in [0,1],$$

where $\otimes$ stands for the element-wise multiplication of vectors. By replacing $A1$ with $A1'$, P is not restricted to be concave (e.g., the underlying production function can be S-shaped) and is capable of capturing the potential existence of increasing, constant, or decreasing marginal products. For empirical studies that use a non-concave production technology, see [10, 11], among others.

In this study, we proceed with the performance measurement using P that satisfies closedness, boundedness, geometric convexity, and free disposability. Based on a given P, the generalized multiplicative directional distance function (GMDDF, [1]) is defined as

$$M(\boldsymbol{x}, \boldsymbol{y}; -\boldsymbol{g}^{-}, \boldsymbol{g}^{+}) := \sup\left\{ \left(\prod_{i=1}^{m} \alpha_i\right)^{\frac{1}{m}} \times \left(\prod_{r=1}^{s} \beta_r\right)^{\frac{1}{r}} \middle| \left(\boldsymbol{\alpha}^{-\boldsymbol{g}^{-}} \boldsymbol{x}, \boldsymbol{\beta}^{\boldsymbol{g}^{+}} \boldsymbol{y}\right) \in P, \boldsymbol{\alpha} \geq \boldsymbol{1}_m, \boldsymbol{\beta} \geq \boldsymbol{1}_s \right\}, \tag{1}$$

where $\boldsymbol{g} := (-\boldsymbol{g}^{-}, \boldsymbol{g}^{+}) \in (-\mathbb{R}_{+}^{m}) \times \mathbb{R}_{+}^{s}$ is a prespecified direction vector and

$$\boldsymbol{\alpha}^{-\boldsymbol{g}^{-}} := \begin{bmatrix} \alpha_1^{-g_1^{-}} & \cdots & 0 \\ \vdots & \ddots & \vdots \\ 0 & \cdots & \alpha_m^{-g_m^{-}} \end{bmatrix} \text{ and } \boldsymbol{\beta}^{\boldsymbol{g}^{+}} := \begin{bmatrix} \beta_1^{g_1^{+}} & \cdots & 0 \\ \vdots & \ddots & \vdots \\ 0 & \cdots & \beta_s^{g_s^{+}} \end{bmatrix}.$$

According to definition (1), both inputs and outputs are changed simultaneously along with a given direction $\boldsymbol{g}$ and are eventually projected onto the frontier (boundary) of P. Denote the points on the frontier of P as $(\boldsymbol{x}^*, \boldsymbol{y}^*)$. We hereby list the main properties of $M(\boldsymbol{x}, \boldsymbol{y}; -\boldsymbol{g}^{-}, \boldsymbol{g}^{+})$ used in this study, as below (See [1] for details on the proofs):

P1: $M(\boldsymbol{x}, \boldsymbol{y}; -\boldsymbol{g}^{-}, \boldsymbol{g}^{+}) \geq 1$ if and only if $(\boldsymbol{x}, \boldsymbol{y}) \in P$.

P2: $M(\boldsymbol{x}^*, \boldsymbol{y}^*; -\boldsymbol{g}^{-}, \boldsymbol{g}^{+}) = 1$ for any given $(-\boldsymbol{g}^{-}, \boldsymbol{g}^{+}) \in (-\mathbb{R}_{+}^{m}) \times \mathbb{R}_{+}^{s}$.

P3: $\log M(\boldsymbol{x}, \boldsymbol{y}; -\boldsymbol{g}^{-}, \boldsymbol{g}^{+})$ is concave on $\log P$,
where $\log P := \{(\log \boldsymbol{x}, \log \boldsymbol{y}) \in \mathbb{R}^m \times \mathbb{R}^s | (\boldsymbol{x}, \boldsymbol{y}) \in P\}$.

The property P1 means that $M(\boldsymbol{x}, \boldsymbol{y}; -\boldsymbol{g}^{-}, \boldsymbol{g}^{+})$ is a functional representation of P, that is

$$P = \{(\boldsymbol{x}, \boldsymbol{y}) \in \mathbb{R}_{++}^{m} \times \mathbb{R}_{++}^{s} | M(\boldsymbol{x}, \boldsymbol{y}; -\boldsymbol{g}^{-}, \boldsymbol{g}^{+}) \geq 1\} \text{ for any given} (-\boldsymbol{g}^{-}, \boldsymbol{g}^{+}) \in (-\mathbb{R}_{+}^{m}) \times \mathbb{R}_{+}^{s}.$$

We first consider a deterministic scenario, in which all producible input-output vectors are only affected by inefficiencies.

Proposition 1.1. Let $\delta_{x_i} \leq 1$ for all $i = 1, \ldots, m$ and $\delta_{y_r} \geq 1$ for all $r = 1, \ldots, s$. Assume that a projection point $(\boldsymbol{x}^*, \boldsymbol{y}^*)$ for the DMU $(\boldsymbol{x}, \boldsymbol{y})$ is achieved by improving $(\boldsymbol{x}, \boldsymbol{y})$ at exponential changes of a vector $\left(\boldsymbol{\delta}_x, \frac{1}{\boldsymbol{\delta}_y}\right)$ by the power of a given vector $(\boldsymbol{g}^-, -\boldsymbol{g}^+) \in \mathbb{R}_+^m \times (-\mathbb{R}_+^s)$. Specifically,

$$(\boldsymbol{x}^*, \boldsymbol{y}^*) = \left(\boldsymbol{\delta}_x^{\boldsymbol{g}^-} \boldsymbol{x}, \left(\frac{1}{\boldsymbol{\delta}_y}\right)^{-\boldsymbol{g}^+} \boldsymbol{y}\right), \tag{2}$$

where

$$\boldsymbol{\delta}_x^{\boldsymbol{g}^-} := \begin{bmatrix} \delta_{x_1}^{g_1^-} & \cdots & 0 \\ \vdots & \ddots & \vdots \\ 0 & \cdots & \delta_{x_m}^{g_m^-} \end{bmatrix} \text{ and } \left(\frac{1}{\boldsymbol{\delta}_y}\right)^{-\boldsymbol{g}^+} := \begin{bmatrix} \left(\frac{1}{\delta_{y_1}}\right)^{-g_1^+} & \cdots & 0 \\ \vdots & \ddots & \vdots \\ 0 & \cdots & \left(\frac{1}{\delta_{y_s}}\right)^{-g_s^+} \end{bmatrix}.$$

Then, the value of $M(\boldsymbol{x}, \boldsymbol{y}; -\boldsymbol{g}^-, \boldsymbol{g}^+)$ for the observed DMU $(\boldsymbol{x}, \boldsymbol{y})$ satisifies

$$M(\boldsymbol{x}, \boldsymbol{y}; -\boldsymbol{g}^-, \boldsymbol{g}^+) = u, \tag{3}$$

where

$$u := \left(\prod_{i=1}^{m} \frac{1}{\delta_{x_i}}\right)^{\frac{1}{m}} \left(\prod_{r=1}^{s} \delta_{y_r}\right)^{\frac{1}{s}}. \tag{4}$$

Proof: According to the definition (1), we have

$$\begin{aligned}
&M\left(\boldsymbol{\delta}_x^{\boldsymbol{g}^-} \boldsymbol{x}, \left(\frac{1}{\boldsymbol{\delta}_y}\right)^{-\boldsymbol{g}^+} \boldsymbol{y}; -\boldsymbol{g}^-, \boldsymbol{g}^+\right) \\
&= \sup\left\{\left(\prod_{i=1}^{m} \alpha_i\right)^{\frac{1}{m}} \times \left(\prod_{r=1}^{s} \beta_r\right)^{\frac{1}{r}} \middle| \left(\boldsymbol{\alpha}^{-\boldsymbol{g}^-} \otimes \boldsymbol{\delta}_x^{\boldsymbol{g}^-} \boldsymbol{x}, \boldsymbol{\beta}^{\boldsymbol{g}^+} \otimes \left(\frac{1}{\boldsymbol{\delta}_y}\right)^{-\boldsymbol{g}^+} \boldsymbol{y}\right) \in P, \boldsymbol{\alpha} \geq \mathbf{1}_m, \boldsymbol{\beta} \geq \mathbf{1}_s\right\} \\
&= \left(\prod_{i=1}^{m} \delta_{x_i}\right)^{\frac{1}{m}} \left(\prod_{r=1}^{s} \frac{1}{\delta_{y_r}}\right)^{\frac{1}{s}} \sup\left\{\left(\prod_{i=1}^{m} \frac{\alpha_i}{\delta_{x_i}}\right)^{\frac{1}{m}} \times \left(\prod_{r=1}^{s} \beta_r \delta_{y_r}\right)^{\frac{1}{r}} \middle| \begin{array}{l} \left(\boldsymbol{\alpha}^{-\boldsymbol{g}^-} \otimes \boldsymbol{\delta}_x^{\boldsymbol{g}^-} \boldsymbol{x}, \boldsymbol{\beta}^{\boldsymbol{g}^+} \otimes \left(\frac{1}{\boldsymbol{\delta}_y}\right)^{-\boldsymbol{g}^+} \boldsymbol{y}\right) \in P, \\ \boldsymbol{\alpha} \oslash \boldsymbol{\delta}_x \geq \mathbf{1}_m, \boldsymbol{\beta} \otimes \boldsymbol{\delta}_y \geq \mathbf{1}_s \end{array}\right\} \\
&= \left(\prod_{i=1}^{m} \delta_{x_i}\right)^{\frac{1}{m}} \left(\prod_{r=1}^{s} \frac{1}{\delta_{y_r}}\right)^{\frac{1}{s}} M(\boldsymbol{x}, \boldsymbol{y}; -\boldsymbol{g}^-, \boldsymbol{g}^+).
\end{aligned}$$

Note that $\oslash$ denotes the element-wise division of vectors. Due to the assumption (2), we obtain

$$\begin{aligned}
M(\boldsymbol{x}^*, \boldsymbol{y}^*; -\boldsymbol{g}^-, \boldsymbol{g}^+) &= M\left(\boldsymbol{\delta}_x^{\boldsymbol{g}^-} \boldsymbol{x}, \left(\frac{1}{\boldsymbol{\delta}_y}\right)^{-\boldsymbol{g}^+} \boldsymbol{y}; -\boldsymbol{g}^-, \boldsymbol{g}^+\right) \\
&= \left(\prod_{i=1}^{m} \delta_{x_i}\right)^{\frac{1}{m}} \left(\prod_{r=1}^{s} \frac{1}{\delta_{y_r}}\right)^{\frac{1}{s}} M(\boldsymbol{x}, \boldsymbol{y}; -\boldsymbol{g}^-, \boldsymbol{g}^+).
\end{aligned}$$

Because of P2, we obtain

$$M(\boldsymbol{x}, \boldsymbol{y}; -\boldsymbol{g}^-, \boldsymbol{g}^+) = \left(\prod_{i=1}^{m} \frac{1}{\delta_{x_i}}\right)^{\frac{1}{m}} \left(\prod_{r=1}^{s} \delta_{y_r}\right)^{\frac{1}{s}}.$$

It is worthnoting that Eq. (4) is equal to the reciprocal of the geometric distance function (GDF) introduced by [2]:

$$\phi := \left(\prod_{i=1}^{m} \delta_{x_i}\right)^{\frac{1}{m}} \left(\prod_{r=1}^{s} \frac{1}{\delta_{y_r}}\right)^{\frac{1}{s}},$$

which is a non-oriented efficiency measure that incorporates both input contraction and output expansion toward the projection points. Note that the original GDF is developed based on the ordinary convexity assumption. However, it is possible to reformulate the GDF on a given P (that satisfies geometric convexity). In such a case, the GDF can be considered a specific case of the GMDDF with $(-\boldsymbol{g}^-, \boldsymbol{g}^+) = (-\mathbf{1}_m, \mathbf{1}_s)$, that is, $M(\boldsymbol{x}, \boldsymbol{y}; -\mathbf{1}_m, \mathbf{1}_s)$. In the scenario described in Proposition 1, the relationship between the GMDDF and GDF can be stated as

$$\begin{aligned} M(\boldsymbol{x}^*, \boldsymbol{y}^*; -\boldsymbol{g}^-, \boldsymbol{g}^+) &= M(\boldsymbol{x}, \boldsymbol{y}; -\mathbf{1}_m, \mathbf{1}_s) \times M(\boldsymbol{x}, \boldsymbol{y}; -\boldsymbol{g}^-, \boldsymbol{g}^+) \\ &= \phi \times M(\boldsymbol{x}, \boldsymbol{y}; -\boldsymbol{g}^-, \boldsymbol{g}^+). \end{aligned}$$

Consequently, $u(= 1/\phi)$ can be used as an inefficiency measure with a range of $[1, \infty)$.

3. A stochastic multiplicative directional distance function

In this section, we further consider a stochastic scenario in which all observations (i.e., $(\boldsymbol{x}_j, \boldsymbol{y}_j), \forall j = 1, \ldots, n$) in P are random variables and have their unique deterministic projection point (efficient point) on the frontier of P (i.e., $(\boldsymbol{x}_j^*, \boldsymbol{y}_j^*)$, $\forall j = 1, \ldots, n$). We assume that the difference between the observation point and the projection point is not only caused by the inefficiency term but also by random noise. Since the generalized multiplicative directional distance function ranges from 1 to ∞, it does not have a straightforward representation to describe the stochastic setting. Therefore, we introduce a *stochatsic* multiplicative directional distance function (SMDDF), which is defined as follows:

$$\tilde{M}(\boldsymbol{x}, \boldsymbol{y}; -\boldsymbol{g}^-, \boldsymbol{g}^+) := \sup\left\{ \left(\prod_{i=1}^{m} \alpha_i\right)^{\frac{1}{m}} \times \left(\prod_{r=1}^{s} \beta_r\right)^{\frac{1}{r}} \middle| \left(\boldsymbol{\alpha}^{-\mathrm{g}^-} \boldsymbol{x}, \boldsymbol{\beta}^{\mathrm{g}^+} \boldsymbol{y}\right) \in \boldsymbol{P}, \boldsymbol{\alpha} > \mathbf{0}_m, \boldsymbol{\beta} > \mathbf{0}_s \right\}, \tag{5}$$

where $\boldsymbol{x}$ and $\boldsymbol{y}$ are random variables.

Some properties of $\tilde{M}(\boldsymbol{x}, \boldsymbol{y}; -\boldsymbol{g}^-, \boldsymbol{g}^+)$ are described as below:

P1': $\tilde{M}(\boldsymbol{x}, \boldsymbol{y}; -\boldsymbol{g}^-, \boldsymbol{g}^+) > 0$ if and only if $(\boldsymbol{x}, \boldsymbol{y}) \in P$.

P2': $\tilde{M}(\boldsymbol{x}^*,\boldsymbol{y}^*;-\boldsymbol{g}^-,\boldsymbol{g}^+) = 1$ for any given $(-\boldsymbol{g}^-,\boldsymbol{g}^+) \in (-\mathbb{R}^m_+) \times \mathbb{R}^s_+$.

P3': $\log\tilde{M}(\boldsymbol{x},\boldsymbol{y};-\boldsymbol{g}^-,\boldsymbol{g}^+)$ is concave on $\log P$,

where $\log P := \{(\log\boldsymbol{x}, \log\boldsymbol{y}) \in \mathbb{R}^m \times \mathbb{R}^s | (\boldsymbol{x},\boldsymbol{y}) \in P\}$.

The property P1' is evident from the definition (5). P2' holds because the frontier containing all of the projection points $(\boldsymbol{x}^*,\boldsymbol{y}^*)$ is assumed to be deterministic. P3' can be proved in a similar way with P3 (see [1] for further details). Furthermore, $\tilde{M}(\boldsymbol{x},\boldsymbol{y};-\boldsymbol{g}^-,\boldsymbol{g}^+)$ has the following translation property:

Proposition 1.2. For any positive vector $(\boldsymbol{\theta},\boldsymbol{\eta}) \in \mathbb{R}^{m+s}_{++}$,

$$\log\tilde{M}\left(\boldsymbol{\theta}^{-\boldsymbol{g}^-}\boldsymbol{x},\boldsymbol{\eta}^{\boldsymbol{g}^+}\boldsymbol{y};-\boldsymbol{g}^-,\boldsymbol{g}^+\right) = \log\tilde{M}(\boldsymbol{x},\boldsymbol{y};-\boldsymbol{g}^-,\boldsymbol{g}^+) - \left(\frac{1}{m}\sum_{i=1}^m \log\theta_i + \frac{1}{s}\sum_{r=1}^s \log\eta_r\right)$$

holds.

Proof: From the definition (5), we obtain

$$\log\tilde{M}(\boldsymbol{x},\boldsymbol{y};-\boldsymbol{g}^-,\boldsymbol{g}^+) = \log\left(\sup\left\{\left(\prod_{i=1}^m \alpha_i\right)^{\frac{1}{m}} \times \left(\prod_{r=1}^s \beta_r\right)^{\frac{1}{r}} \middle| \begin{array}{l}\left(\boldsymbol{\alpha}^{-\boldsymbol{g}^-}\boldsymbol{x},\boldsymbol{\beta}^{\boldsymbol{g}^+}\boldsymbol{y}\right) \in P,\\ \boldsymbol{\alpha} > \boldsymbol{0}_m, \boldsymbol{\beta} > \boldsymbol{0}_s\end{array}\right\}\right)$$

$$= \sup\left\{\frac{1}{m}\sum_{i=1}^m \log\alpha_i + \frac{1}{s}\sum_{r=1}^s \log\beta_r \middle| \begin{array}{l}(\log\boldsymbol{x} - \boldsymbol{g}^- \otimes \log\boldsymbol{\alpha}, \log\boldsymbol{y} + \boldsymbol{g}^+ \otimes \log\boldsymbol{\beta}) \in \log P,\\ \log\boldsymbol{\alpha}: \text{ free}, \log\boldsymbol{\beta}: \text{ free}\end{array}\right\}.$$

Therefore, for any positive vector $(\boldsymbol{\theta},\boldsymbol{\eta}) \in \mathbb{R}^{m+s}_{++}$, we obtain

$$\log\tilde{M}\left(\boldsymbol{\theta}^{-\boldsymbol{g}^-}\boldsymbol{x},\boldsymbol{\eta}^{\boldsymbol{g}^+}\boldsymbol{y};-\boldsymbol{g}^-,\boldsymbol{g}^+\right) = \sup\left\{\frac{1}{m}\sum_{i=1}^m \log\alpha_i + \frac{1}{s}\sum_{r=1}^s \log\beta_r \middle| \right.$$

$$\left.\begin{array}{l}(\log\boldsymbol{x} - \boldsymbol{g}^- \otimes \log\boldsymbol{\theta} - \boldsymbol{g}^- \otimes \log\boldsymbol{\alpha}, \log\boldsymbol{y} + \boldsymbol{g}^+ \otimes \log\boldsymbol{\eta} + \boldsymbol{g}^+ \otimes \log\boldsymbol{\beta}) \in \log P,\\ \log\boldsymbol{\alpha}: \text{ free}, \log\boldsymbol{\beta}: \text{ free}\end{array}\right\}$$

$$= -\left(\frac{1}{m}\sum_{i=1}^m \log\theta_i + \frac{1}{s}\sum_{r=1}^s \log\eta_r\right)$$

$$+\sup\left\{\frac{1}{m}\sum_{i=1}^m \log\alpha_i + \frac{1}{s}\sum_{r=1}^s \log\beta_r + \left(\frac{1}{m}\sum_{i=1}^m \log\theta_i + \frac{1}{s}\sum_{r=1}^s \log\eta_r\right) \middle| \right.$$

$$\left.\begin{array}{l}(\log\boldsymbol{x} - \boldsymbol{g}^- \otimes (\log\boldsymbol{\alpha} + \log\boldsymbol{\theta}), \log\boldsymbol{y} + \boldsymbol{g}^+ \otimes (\log\boldsymbol{\beta} + \log\boldsymbol{\eta})) \in \log P,\\ \log\boldsymbol{\alpha} + \log\boldsymbol{\theta}: \text{ free}, \log\boldsymbol{\beta} + \log\boldsymbol{\eta}: \text{ free}\end{array}\right\}$$

$$= \log\tilde{M}(\boldsymbol{x},\boldsymbol{y};-\boldsymbol{g}^-,\boldsymbol{g}^+) - \left(\frac{1}{m}\sum_{i=1}^m \log\theta_i + \frac{1}{s}\sum_{r=1}^s \log\eta_r\right).$$

It is worth noting that the proposed $\tilde{M}(\boldsymbol{x},\boldsymbol{y};-\boldsymbol{g}^-,\boldsymbol{g}^+)$ precisely reveals a multiplicative data-generating process. To help inform intuition regarding this type of data-generating process, let $\epsilon_{x_i} > 0$ for all $i = 1, \ldots, m$ and $\epsilon_{y_r} > 0$ for all $r = 1, \ldots, s$ be the element-specific composite errors for inputs and outputs, respectively. Since $\boldsymbol{x}$ and $\boldsymbol{y}$ are random variables that are subject to both inefficiencies and noises, we assume that $E(\epsilon_{x_i}) = \mu_{x_i} > 0$, $Var(\epsilon_{x_i}) < \infty$ and $E(\epsilon_{y_i}) = \mu_{y_i} > 0$, $Var(\epsilon_{y_i}) < \infty$. Suppose that

$$(\boldsymbol{x}, \boldsymbol{y}) = \left(\epsilon_{\boldsymbol{x}}^{\boldsymbol{g}^-} \boldsymbol{x}^*, \epsilon_{\boldsymbol{y}}^{-\boldsymbol{g}^+} \boldsymbol{y}^*\right). \tag{6}$$

Then, it follows from Proposition 1.2 and property P2' that

$$\log \tilde{M}(\boldsymbol{x}, \boldsymbol{y}; -\boldsymbol{g}^-, \boldsymbol{g}^+) = \frac{1}{m}\sum_{i=1}^{m} \log \epsilon_{x_i} + \frac{1}{s}\sum_{r=1}^{s} \log \epsilon_{y_r}, \tag{7}$$

which is equivalent to

$$\tilde{M}(\boldsymbol{x}, \boldsymbol{y}; -\boldsymbol{g}^-, \boldsymbol{g}^+) = \left(\prod_{i=1}^{m} \epsilon_{x_i}\right)^{\frac{1}{m}} \times \left(\prod_{r=1}^{s} \epsilon_{y_r}\right)^{\frac{1}{r}}.$$

To simplify the notation, we define $\epsilon := \left(\prod_{i=1}^{m} \epsilon_{x_i}\right)^{\frac{1}{m}} \times \left(\prod_{r=1}^{s} \epsilon_{y_r}\right)^{\frac{1}{r}}$. Then,

$$\tilde{M}(\boldsymbol{x}, \boldsymbol{y}; -\boldsymbol{g}^-, \boldsymbol{g}^+) = \epsilon.$$

Therefore, $\tilde{M}(\boldsymbol{x}, \boldsymbol{y}; -\boldsymbol{g}^-, \boldsymbol{g}^+)$ can be interpreted as a composite error term that is constructed with ϵ_{x_i} $(\forall i)$ and ϵ_{y_r} $(\forall r)$. Moreover, from the linearity of expectation and Jensen's inequality, we also have $0 < \mu := E(\log \epsilon) = E\left(\frac{1}{m}\sum_{i=1}^{m} \log \epsilon_{x_i} + \frac{1}{s}\sum_{r=1}^{s} \log \epsilon_{y_r}\right) \leq \frac{1}{m}\sum_{i=1}^{m} \log \mu_{x_i} + \frac{1}{s}\sum_{r=1}^{s} \log \mu_{y_r}$.

To estimate the inefficiency for the observed DMU$_j$ ($j = 1, \ldots, n$) as well as the SMDDF, we apply Proposition 1.2 and set $\left(\boldsymbol{\theta}_j, \boldsymbol{\eta}_j\right) = \left(\boldsymbol{x}_j, \boldsymbol{y}_j\right)$ to obtain the following equality:

$$\log \tilde{M}\left(\boldsymbol{x}_j^{\boldsymbol{g}^-} \boldsymbol{x}_j, \boldsymbol{y}_j^{-\boldsymbol{g}^+} \boldsymbol{y}_j; -\boldsymbol{g}^-, \boldsymbol{g}^+\right) = \log \tilde{M}\left(\boldsymbol{x}_j, \boldsymbol{y}_j; -\boldsymbol{g}^-, \boldsymbol{g}^+\right) + \left(\frac{1}{m}\sum_{i=1}^{m} \log x_{ij} + \frac{1}{s}\sum_{r=1}^{s} \log y_{rj}\right).$$

Using Eq. (7), we further obtain

$$\begin{aligned}\frac{1}{m}\sum_{i=1}^{m} \log x_{ij} + \frac{1}{s}\sum_{r=1}^{s} \log y_{rj} &= \log \tilde{M}\left(\boldsymbol{x}_j^{\boldsymbol{g}^-} \boldsymbol{x}_j, \boldsymbol{y}_j^{-\boldsymbol{g}^+} \boldsymbol{y}_j; -\boldsymbol{g}^-, \boldsymbol{g}^+\right) - \log \tilde{M}\left(\boldsymbol{x}_j, \boldsymbol{y}_j; -\boldsymbol{g}^-, \boldsymbol{g}^+\right) \\ &= \log \tilde{M}\left(\boldsymbol{x}_j^{\boldsymbol{g}^-} \boldsymbol{x}_j, \boldsymbol{y}_j^{-\boldsymbol{g}^+} \boldsymbol{y}_j; -\boldsymbol{g}^-, \boldsymbol{g}^+\right) - \log \epsilon_j.\end{aligned}$$

Let $\tilde{\boldsymbol{x}}_j = (\boldsymbol{x}_j)^{\boldsymbol{g}^-} \boldsymbol{x}_j$ and $\tilde{\boldsymbol{y}}_j = \left(\boldsymbol{y}_j\right)^{-\boldsymbol{g}^+} \boldsymbol{y}_j$. Then,

$$\begin{aligned}\overline{\overline{M}}\left(\tilde{\boldsymbol{x}}_j, \tilde{\boldsymbol{y}}_j; -\boldsymbol{g}^-, \boldsymbol{g}^+\right) &:= E\left(\log \tilde{M}\left(\tilde{\boldsymbol{x}}_j, \tilde{\boldsymbol{y}}_j; -\boldsymbol{g}^-, \boldsymbol{g}^+\right)\right) - E\left(\log \epsilon_j\right) \\ &= \log \tilde{M}\left(\tilde{\boldsymbol{x}}_j, \tilde{\boldsymbol{y}}_j; -\boldsymbol{g}^-, \boldsymbol{g}^+\right) - \mu\end{aligned}$$

It follows from P3 that $\tilde{M}\left(\tilde{\boldsymbol{x}}_j, \tilde{\boldsymbol{y}}_j; -\boldsymbol{g}^-, \boldsymbol{g}^+\right)$ is a concave function. Moreover, $\overline{\overline{M}}\left(\tilde{\boldsymbol{x}}_j, \tilde{\boldsymbol{y}}_j; -\boldsymbol{g}^-, \boldsymbol{g}^+\right)$ could be seen as the conditional expectation function of $\left(\frac{1}{m}\sum_{i=1}^{m} \log x_{ij} + \frac{1}{s}\sum_{r=1}^{s} \log y_{rj}\right)$. That is,

$$\frac{1}{m}\sum_{i=1}^{m}\log x_{ij}+\frac{1}{s}\sum_{r=1}^{s}\log y_{rj}=\overline{\overline{M}}\left(\tilde{\boldsymbol{x}}_j,\tilde{\boldsymbol{y}}_j;-\boldsymbol{g}^-,\boldsymbol{g}^+\right)-\xi_j,$$

where $\xi_j = \log u_j + v_j - \mu$ and u_j is a DMU-specific inefficiency term that satisfies $E(\log u_j) = \mu$, $Var(\log u_j) < \infty$, and v_j is a noise term that satisfies $E(v_j) = 0$, $Var(v_j) < \infty$.

To estimate $\overline{\overline{M}}\left(\tilde{\boldsymbol{x}}_j,\tilde{\boldsymbol{y}}_j;-\boldsymbol{g}^-,\boldsymbol{g}^+\right)$ as well as the average level of inefficiencies, we apply the method of convex nonparametric least squares (CNLS, [6]), which is a nonparametric regression approach with a least-square objective subject to monotonicity and convexity/concavity constraints. However, the definition of SMDDF requires additional conditions that Proposition 1.2 should hold, contrary to constraints of the ordinary CNLS problem. Therefore, we formulate the CNLS problem as the following quadratic programming (QP) problem:

$$\min \sum_{j=1}^{n}\xi_j^2 \tag{8}$$

$$\text{s.t.}\quad \frac{1}{m}\sum_{i=1}^{m}\log x_{ij}+\frac{1}{s}\sum_{r=1}^{s}\log y_{rj}=v_j+\sum_{i=1}^{m}\zeta_{ij}\log\tilde{x}_{ij}-\sum_{r=1}^{s}\gamma_{rj}\log\tilde{y}_{rj}-\xi_j,\ \ \forall j \tag{9}$$

$$v_j+\sum_{i=1}^{m}\zeta_{ij}\log\tilde{x}_{ij}-\sum_{r=1}^{s}\gamma_{rj}\log\tilde{y}_{rj}\le v_h+\sum_{i=1}^{m}\zeta_{ih}\log\tilde{x}_{ij}-\sum_{r=1}^{s}\gamma_{rh}\log\tilde{y}_{rj},\ \ \forall j,h \ \text{ and } \ j\ne h \tag{10}$$

$$\sum_{i=1}^{m}\zeta_{ij}g_{ij}^-\log x_{ij}+\sum_{r=1}^{s}\gamma_{rj}g_{rj}^+\log y_{rj}=\frac{1}{m}\sum_{i=1}^{m}\log x_{ij}+\frac{1}{s}\sum_{r=1}^{s}\log y_{rj},\ \ \forall j \tag{11}$$

$$\zeta_{ij}\ge 0,\gamma_{rj}\ge 0,\forall i,r,j. \tag{12}$$

The objective function (8) and the constraint (9) fit a linear equation for each DMU$_j$, where $\frac{1}{m}\sum_{i=1}^{m}\log x_{ij}+\frac{1}{s}\sum_{r=1}^{s}\log y_{rj}$ is the dependent variable and $\left(\tilde{\boldsymbol{x}}_j,-\tilde{\boldsymbol{y}}_j\right)$ is the independent variable. The second constraint (10) further imposes Afrait's theorem [12], which ensures global concavity of the estimated $\overline{\overline{M}}\left(\tilde{\boldsymbol{x}}_j,\tilde{\boldsymbol{y}}_j;-\boldsymbol{g}^-,\boldsymbol{g}^+\right)$. The third constraint (11) ensures the translation property as described in Proposition 1.2. The last two constraints in (12) impose monotonicity relative to all transformed inputs $\tilde{\boldsymbol{x}}_j$ and outputs $\tilde{\boldsymbol{y}}_j$. Using the translation property, the above problem can be reformulated as

$$\min \sum_{j=1}^{n}\xi_j^2 \tag{13}$$

$$\text{s.t.}\quad \sum_{r=1}^{s}\gamma_{rj}\log y_{rj}=v_j+\sum_{i=1}^{m}\zeta_{ij}\log x_{ij}-\xi_j,\ \ \forall j \tag{14}$$

$$v_j+\sum_{i=1}^{m}\zeta_{ij}\log x_{ij}-\sum_{r=1}^{s}\gamma_{rj}\log y_{rj}\le v_h+\sum_{i=1}^{m}\zeta_{ih}\log x_{ij}-\sum_{r=1}^{s}\gamma_{rh}\log y_{rj},\ \ \forall j,h \ \text{ and } \ j\ne h \tag{15}$$

$$\sum_{i=1}^{m} \zeta_{ij} g_{ij}^{-} \log x_{ij} + \sum_{r=1}^{s} \gamma_{rj} g_{rj}^{+} \log y_{rj} = \frac{1}{m} \sum_{i=1}^{m} \log x_{ij} + \frac{1}{s} \sum_{r=1}^{s} \log y_{rj}, \ \forall j \tag{16}$$

$$\zeta_{ij} \geq 0, \gamma_{rj} \geq 0, \forall i, r, j, \tag{17}$$

where the constraint (14) now is a linear equation of inputs and outputs for each DMU_j. Since the problems (13)–(17) estimate $\overline{\overline{M}}\left(\tilde{\boldsymbol{x}}_j, \tilde{\boldsymbol{y}}_j; -\boldsymbol{g}^-, \boldsymbol{g}^+\right)$ as a concave function, the coefficients $\upsilon_j, \zeta_{ij}, \gamma_{rj}, \forall i, r$ characterize a tangent hyperplane of the estimated $\overline{\overline{M}}\left(\tilde{\boldsymbol{x}}_j, \tilde{\boldsymbol{y}}_j; -\boldsymbol{g}^-, \boldsymbol{g}^+\right)$ for each DMU_j.

After solving the problem (13)–(17), it is possible to derive the expected value of the logarithm of inefficiencies (i.e., μ) based on the skewness of the estimated residuals. For example, μ can be measured using the method of moments [5] (Note that alternative approaches such as quasi-likelihood estimation and nonparametric kernel deconvolution are available. See, e.g., [13] for a detailed discussion).

4. Application to Japanese life insurance companies

In this section, we investigate the productive performance of Japanese life insurance companies by considering both deterministic and stochastic scenarios. The sample consisted of 41 member companies of the Life Insurance Association of Japan (LIAJ). We considered the sample years 2011–2020. The data were drawn from the financial statements of all companies or the *Summary of Life Insurance Business 2021* as published by LIAJ.

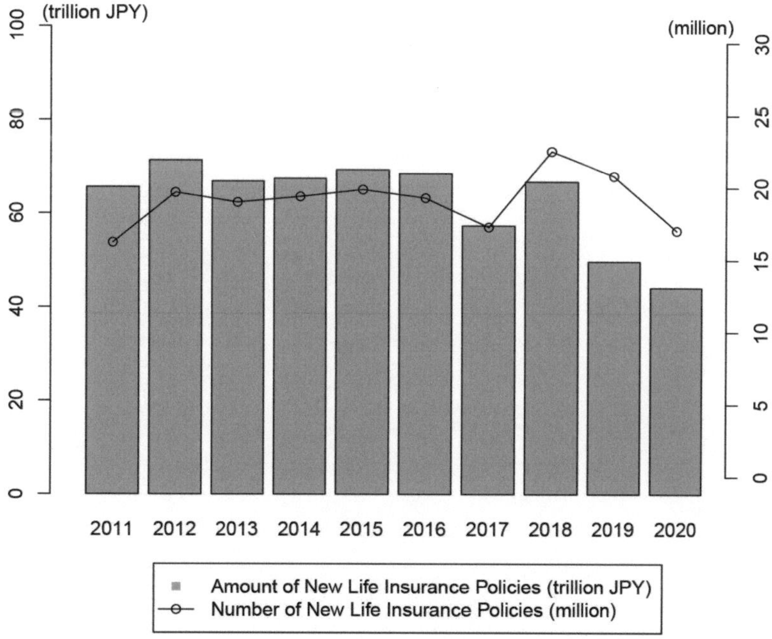

Figure 1.
Trends in individual life insurance business (source: Life Insurance Association of Japan (LIAJ)).

Life insurance companies play an essential role in the financial and loan markets in Japan [14]. Many of them have relied predominantly on face-to-face sales for a long period. However, since the beginning of the COVID-19 pandemic, Japanese life insurance companies were forced to temporarily suspend their face-to-face sales channels due to the nationwide declaration of a state of emergency. As shown in **Figure 1**, domestic sales of new policies have declined significantly since (fiscal year) 2018. According to the *Life Insurance Factbook 2019* published by LIAJ, the decrease in 2019 was primarily due to lower sales of foreign currency-denominated insurance. In 2020, the conditions of domestic sales in new policies became even worse because insurance agents could not meet new customers directly. According to the monthly statistics of LIAJ, the number of new businesses reduced by 34.4% in April and 68.5% in May, on a year-on-year basis during the period of the first declaration of a state of emergency from 2020/4/7 to 2020/5/25. Moreover, even though Japanese life insurance companies resumed face-to-face sales after the state of emergency was lifted, the year-on-year decline in 2020 continued until August (14.3%).

Japanese life insurance companies have changed their distribution channels over the past 10 years. It can be seen from **Figure 2** that the numbers of individual and corporate agencies have been decreasing gradually in recent years. This trend should not be taken as proof that the scale of the life insurance industry has shrunk. In fact, Japanese life insurance companies have become more diversified in their sales channels due to diversifying insurance products and consumer needs and a recently increasing presence of non-traditional sales channels, such as Internet sales. Considering that life insurance companies must refrain from face-to-face sales to prevent the spread of COVID-19, a significant diversion of sales channels may lead to better performance, such as shifting from the traditional business model of face-to-face sales to online sales channels. However, sales representatives in agencies remained the primary sales channel of life insurance in 2020 [15].

Japanese life insurance companies also play a critical role as stable institutional investors in the capital market as they hold a certain level of listed shares for a long period. **Figure 3** shows the breakdown of listed shares by type of shareholder. It can be seen that the amount of listed shares held by insurance companies (insurance and pension funds) has been increasing over the past 10 years. Thus, measuring the behavior of insurance companies makes economic sense. In addition, it has been widely recognized that the current solvency margin ratio regulation introduced in 1996 does

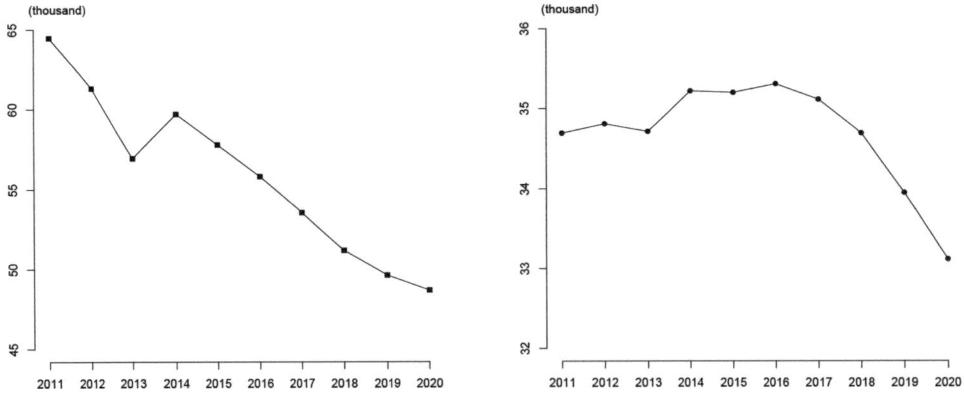

Figure 2.
Distribution channels (source: Life Insurance Association of Japan (LIAJ)).

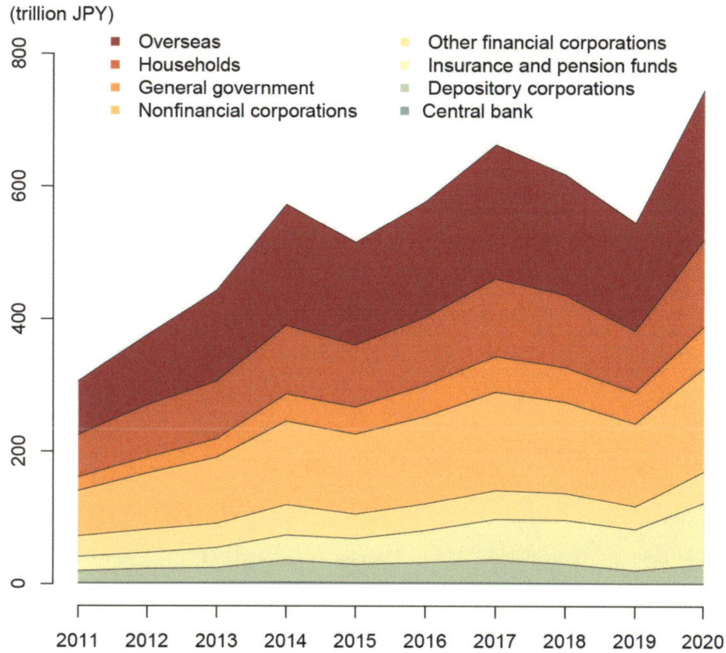

Figure 3.
Breakdown of listed shares by type of shareholder (source: Bank of Japan).

not adequately assess the valuation of liabilities and the amount of risk. This issue is expected to be solved by introducing economic value-based solvency regulations. According to the report on the economic value-based solvency framework released by the Financial Services Agency (FSA) of Japan in June 2020, the economic value-based solvency regulation will be enforced from April 2025. Since new regulations may change the business environment of Japanese life insurance companies and affect their behavior, it makes practical sense to study the past and present productive performance of Japanese life insurance companies to better prepare for the arrival of new regulations.

4.1 Inputs and outputs

To characterize the inputs and outputs of Japanese life insurance companies, we applied a value-added approach that has been commonly used in previous studies (see, e.g., [16] for a detailed discussion of the value-added approach). We considered three inputs and two outputs. The inputs were "office workers (x_1)," "total shareholders' equity (x_2)," and "operating expenses (x_3)." The first two variables, office workers and total shareholders' equity, were proxy variables for labor and capital, respectively. The operating expenses included various cost items required to operate the life insurance business, such as the selling cost, management/maintenance-related cost, and others. In contrast, the outputs were "ordinary income (y_1)" and "total payment of insurance benefits (y_2)." Ordinary predominately derives from the insurance business, such as premium income and investment-related business, the latter of which includes interest and dividend income. To account for the risk-pooling and risk-bearing capabilities of Japanese life insurance companies, we also considered the total payment of insurance benefits as a proxy output variable.

The descriptive statistics of inputs and outputs are summarized in **Table 1**.

Years	Statistics[a]	x_1	x_2	x_3	y_1	y_2
2020						
	mean	2780.76	3036.88	1118.93	7350.61	11960.23
	st. dev.	4148.45	4580.88	1371.91	11467.07	15925.40
	min	58.00	7.48	18.41	25.09	86.06
	max	21117.00	18952.08	5752.23	58660.91	67862.10
2019						
	mean	2727.00	2799.84	1147.76	7206.73	11418.84
	st. dev.	3952.85	4297.92	1429.95	11701.79	16192.90
	min	31.00	12.68	9.66	21.10	61.08
	max	19425.00	17544.31	5987.46	61913.69	72114.03
2018						
	mean	2708.49	2873.19	1190.10	7021.15	11835.05
	st. dev.	3950.21	4343.64	1463.74	12403.06	17165.93
	min	31.00	17.49	9.59	17.43	50.25
	max	19392.00	17851.78	6119.73	68688.93	79166.54
2017						
	mean	2659.93	2754.62	1158.89	7059.29	11528.79
	st. dev.	3969.84	4287.69	1478.26	12414.73	16958.81
	min	34.00	9.21	9.88	13.18	42.84
	max	19515.00	18012.51	6005.71	68900.20	79529.49
2016						
	mean	2643.37	2605.89	1152.84	7076.60	11649.00
	st. dev.	4000.73	4174.32	1486.14	13265.85	18130.89
	min	33.00	10.69	9.29	12.90	36.72
	max	19747.00	17412.86	5721.59	75503.23	86594.42

[a] *Units of x_2, x_3, y_1, y_2: trillion yen; the unit of x_1: person.*

Table 1.
Descriptive statistics of inputs and outputs.

4.2 Deterministic scenarios

In previous studies, the geometric distance function (GDF) was measured *a posteriori* after the projection points were computed due to its nonlinearity [2, 17]. In Section 2, we showed that for a certain deterministic scenario (see, Proposition 1.1), the value of generalized multiplicative directional distance function $M(\boldsymbol{x},\boldsymbol{y};-\boldsymbol{g}^-,\boldsymbol{g}^+)$ is equal to the reciprocal of GDF. In this section, we calculated the GDP for each Japanese insurance company using the function $M(\boldsymbol{x},\boldsymbol{y};-\boldsymbol{g}^-,\boldsymbol{g}^+)$.

For any $\left(\frac{1}{\boldsymbol{\delta}_x},\boldsymbol{\delta}_y\right) \geq (\mathbf{1}_m,\mathbf{1}_s)$, the function $M(\boldsymbol{x},\boldsymbol{y};-\boldsymbol{g}^-,\boldsymbol{g}^+)$ can be measured by the following nonlinear programming (NLP) problem:

$$
\begin{aligned}
\max \quad & \left(\prod_{i=1}^{m} \frac{1}{\delta_{x_i}}\right)^{\frac{1}{m}} \left(\prod_{r=1}^{s} \delta_{y_r}\right)^{\frac{1}{s}} \\
\text{s.t.} \quad & \prod_{j=1}^{n} x_{ij}^{\lambda_j} \le \left(\frac{1}{\delta_{x_i}}\right)^{-g_i^-} x_{io}, \quad i = 1, \dots, m \\
& \prod_{j=1}^{n} y_{rj}^{\lambda_j} \ge \delta_{y_r}^{g_r^+} y_{ro}, \quad r = 1, \dots, s \\
& \sum_{j=1}^{n} \lambda_j = 1 \\
& 0 < \delta_{x_i} \le 1, \delta_{y_r} \ge 1, \lambda_j \ge 0, \quad \forall i, r, j.
\end{aligned}
$$

Because of P3, the above problem can be equivalently reformulated as

$$
\begin{aligned}
\max \quad & \frac{1}{s}\sum_{j=1}^{n} \log \delta_{y_r} - \frac{1}{m}\sum_{j=1}^{n} \log \delta_{x_i} \\
\text{s.t.} \quad & \sum_{j=1}^{n} \lambda_j \log x_{ij} \le \log x_{io} + g_i^- \log \delta_{x_i}, \quad i = 1, \dots, m \\
& \sum_{j=1}^{n} \lambda_j \log y_{rj} \ge \log y_{ro} + g_r^+ \log \delta_{y_r}, \quad r = 1, \dots, s \\
& \sum_{j=1}^{n} \lambda_j = 1 \\
& \log \delta_{x_i} \le 0, \log \delta_{y_r} \ge 0, \lambda_j \ge 0, \quad \forall i, r, j.
\end{aligned}
$$

Let $\left(\delta_{x_i}^*, \delta_{y_r}^*, \lambda_j^*\right), \forall i, r, j$ be an optimal solution to the above model. Then, the GDF can be computed by

$$
\phi^* = \left(\prod_{i=1}^{m} \delta_{x_i}^*\right)^{\frac{1}{m}} \left(\prod_{r=1}^{s} \frac{1}{\delta_{y_r}^*}\right)^{\frac{1}{s}}.
$$

From a theoretical perspective, an exogenously given direction vector $(\boldsymbol{g}^-, \boldsymbol{g}^+)$ is preferred for the determination of projection points. However, from a computational perspective, the direction vector in the above model can be chosen arbitrarily. In practice, the decision-makers may choose proper direction vectors depending on their purpose of performance benchmarking, or alternatively, use the direction vectors that lead to unit-invariant efficiency measures (See, e.g., [1] for detailed discussion on selecting direction vectors). In this study, we adopted $\left(g_i^-, g_r^+\right) = \left(\log x_{io} - \min_{j \in \{1, \dots, n\}} \log x_{ij}, \max_{j \in \{1, \dots, j\}} \log y_{rj} - \log y_{ro}\right), \forall i, r$ to derive the unit-invariant GDF efficiencies.

The estimated GDF efficiencies are reported in **Table 2**. There were three efficient companies over the fiscal years 2016–2020: DMU05 (AEON Allianz Life Insurance Co., Ltd.), DMU18 (The Dai-ichi Frontier Life Insurance Co., Ltd.), and DMU41

DMUs[a]	2020	2019	2018	2017	2016
DMU01	0.501	0.486	0.6	0.61	0.677
DMU02	1	1	1	0.3	0.3
DMU03	0.433	0.449	0.557	0.538	0.588
DMU04	0.437	0.44	0.504	0.519	0.557
DMU05	1	1	1	1	1
DMU06	0.438	0.418	0.342	0.471	0.759
DMU07	0.638	0.619	0.966	1	0.451
DMU08	0.618	0.555	0.692	0.815	0.876
DMU09	0.475	0.45	0.57	0.616	0.705
DMU10	0.486	0.551	0.448	0.621	0.628
DMU11	1	0.779	0.38	1	1
DMU12	0.512	0.516	0.652	0.628	0.673
DMU13	0.467	0.446	0.52	0.504	0.548
DMU14	0.5	0.496	0.584	0.567	0.6
DMU15	0.591	1	0.621	0.828	1
DMU16	0.453	0.456	0.582	0.574	0.602
DMU17	0.594	0.531	0.659	0.635	0.728
DMU18	1	1	1	1	1
DMU19	0.472	0.485	0.604	0.599	0.637
DMU20	0.468	0.482	0.598	0.57	0.618
DMU21	0.492	0.501	0.656	0.61	0.684
DMU22	1	1	1	0.6	0.613
DMU23	0.736	0.672	0.616	0.767	1
DMU24	0.493	0.506	0.66	0.674	0.749
DMU25	0.662	0.668	0.74	0.343	0.344
DMU26	0.336	0.343	0.343	0.382	0.303
DMU27	0.587	1	0.544	0.589	0.641
DMU28	0.462	0.479	0.604	0.624	0.661
DMU29	0.811	0.709	0.601	0.696	0.759
DMU30	0.554	0.558	0.722	0.777	0.87
DMU31	0.59	0.588	0.738	0.786	0.821
DMU32	0.695	0.605	1	1	1
DMU33	0.447	0.46	0.566	0.543	0.584
DMU34	1	0.708	0.907	1	0.926
DMU35	0.401	0.398	0.305	0.313	0.309
DMU36	0.396	0.423	0.491	0.492	0.523
DMU37	0.553	0.55	0.716	0.663	0.824
DMU38	0.413	0.45	0.346	0.419	0.4

DMUs[a]	2020	2019	2018	2017	2016
DMU39	0.404	0.438	0.309	0.324	0.321
DMU40	0.419	0.456	0.421	0.539	0.5
DMU41	1	1	1	1	1
mean	0.598	0.602	0.638	0.647	0.678
st. dev	0.208	0.204	0.208	0.206	0.215

[a]*The number of DMUs follows the order of the member companies of the Life Insurance Association of Japan (LIAJ), as released on November 2, 2021. We excluded HANASAKU LIFE INSURANCE Co., Ltd. from the sample because it is a new member as of April 1, 2019.*

Table 2.
Results of GDF efficiencies.

(Rakuten Life Insurance Co., Ltd.). These companies can be seen as adopting best practices that were sustainable given the challenges of a changing business environment over the whole period. In addition, we also found that the Japanese life insurance industry is experiencing a downturn in performance, as the average efficiency score declined from 2016 to 2020.

The average values of $\left(\delta_x^*, \frac{1}{\delta_y^*}\right)$ are summarized in **Table 3**. The parameters δ_x^* and $\frac{1}{\delta_y^*}$ can be interpreted as the element-specific efficiencies for inputs and outputs, respectively. It can be seen from **Table 3** that the input "total shareholders' equity (x_2)" had the highest average efficiency score versus the other inputs and outputs over the period 2016–2020 (the mean of $\delta_{x_2}^*$ was 0.861). The second highest average efficiency score was achieved by "operating expenses (x_3)" (the mean of $\delta_{x_3}^*$ was 0.808). This indicates that Japanese life insurance companies have good performance in funding-related inputs. By contrast, the input "office workers (x_1)" had the lowest average efficiency score (the mean of $\delta_{x_1}^*$ was 0.67), which implies that Japanese life insurance companies have relatively poor performance in human resources. This may be due to the paradigm shift in the face of an increasing number of new sales channels. Additionally, since Japanese life insurance companies could not meet new customers directly in order to prevent the spread of COVID-19 in 2020, the decreased efficiency score in 2020 compared to 2019 may due to the impact of the nationwide declaration of a state of emergency. However, further investigations are needed regarding this issue.

Year	$\delta_{x_1}^*$	$\delta_{x_2}^*$	$\delta_{x_3}^*$	$\frac{1}{\delta_{y_1}^*}$	$\frac{1}{\delta_{y_2}^*}$
2020	0.667	0.925	0.751	0.661	0.73
2019	0.704	0.867	0.85	0.65	0.699
2018	0.648	0.85	0.803	0.794	0.77
2017	0.651	0.833	0.815	0.782	0.78
2016	0.681	0.828	0.819	0.798	0.821
mean	0.67	0.861	0.808	0.737	0.76

Table 3.
Results of the estimated $\left(\delta_x^*, \frac{1}{\delta_y}^*\right)$.

4.3 Stochastic scenarios

Compared with the generalized multiplicative directional distance function $M(\boldsymbol{x},\boldsymbol{y}; -\boldsymbol{g}^{-},\boldsymbol{g}^{+})$, the proposed *stochastic* multiplicative directional distance function (SMDDF) $\tilde{M}(\boldsymbol{x},\boldsymbol{y}; -\boldsymbol{g}^{-},\boldsymbol{g}^{+})$ is capable of capturing the statistical variation present in the observed data. To facilitate comparison, we applied the same direction vector used in Section 4.2 to estimate the SMDDF. The estimated coefficients characterizing the tangent hyperplanes of the concave SMDDF for all DMUs are summarized in **Table 4**.

From **Table 4**, it is clear that the estimated coefficients exhibited significant variance over the period 2016–2020. However, few DMUs had extremely large or extremely small coefficients, and the estimated coefficients of most of the observed DMUs were in the range 0 to 1. This can be intuitively understood, for example, from the box plot of the intercepts υ^* excluding the extreme values, as shown in **Figure 4**. We also report the DMUs with extreme values of estimated coefficients for each year

Years	Statistics[a]	υ^*	ζ_1^*	ζ_2^*	ζ_3^*	γ_1^*	γ_2^*
2020							
	mean	−595.781	719.704	324.489	858.349	424.36	24.15
	st. dev.	23506.536	4604.907	2075.796	3837.448	2714.835	118.149
	min	−56133.412	0.001	0.015	0	0.001	0
	max	121722.234	29486.327	13291.882	17161.857	17383.801	591.261
	†DMU02	−38155.613	2.357	13291.882	1.082	0.139	0.14
	†DMU05	−56133.412	7.441	0.453	17161.857	0.193	0.193
	†DMU11	−51901.156	29486.327	0.44	1.423	0.19	0.202
	†DMU41	121722.234	0.455	0.166	0.286	17383.801	591.261
2019							
	mean	−191.393	461.661	792.405	200.862	232.605	378.179
	st. dev.	25869.435	2954.279	4945.889	851.021	1486.929	1850.085
	min	−95754.967	0.002	0	0	0.002	0
	max	126825.46	18916.892	30887.494	3610.85	9521.375	9064.062
	†DMU02	−95754.967	1.197	30887.494	0.618	0.125	0.126
	†DMU05	−38947.375	18916.892	0.652	3610.85	0.221	0.22
	†DMU41	126825.46	0.393	0.187	0.241	9521.375	9064.062
2018							
	mean	−1174.219	221.483	822.733	286.871	165.586	173.732
	st. dev.	20711.798	1416.317	5001.333	1281.837	1058.141	932.863
	min	−98572.938	0.002	0	0	0.005	0
	max	81008.126	9069.146	30422.438	5732.792	6775.741	5024.123
	†DMU02	−98572.938	1.303	30422.438	0.712	0.122	0.119
	†DMU05	−30606.077	9069.146	3.194	5732.792	0.222	0.226
	†DMU41	81008.126	0.419	0.2	0.25	6775.741	5024.123

Years	Statistics[a]	υ^*	ζ_1^*	ζ_2^*	ζ_3^*	γ_1^*	γ_2^*
2017							
	mean	2470.9	117.05	286.855	338.048	397.955	277.471
	st. dev.	27075.687	747.405	1742.883	1547.98	2545.768	1412.526
	min	−59913.335	0.004	0	0	0.004	0
	max	161196.848	4786.051	10601.869	7093.989	16301.242	7202.949
	†DMU05	−59913.335	4786.051	10601.869	7093.989	0.262	0.267
	†DMU41	161196.848	0.394	0.172	0.232	16301.242	7202.949
2016							
	mean	1543.176	350.93	112.891	35.068	299.912	113.54
	st. dev.	15406.705	1717.644	694.418	222.124	1641.759	714.694
	min	−29985.44	0	0	0.01	0	0
	max	93237.186	8415.021	4280.92	1422.665	8992.456	4520.656
	†DMU05	−29985.44	8415.021	4280.92	1422.665	0.269	0.279
	†DMU41	93237.186	0.414	0.192	0.244	8992.456	4520.656

[a]*† represents the DMU with extremely large or extremely small coefficients.*

Table 4.
Descriptive statistics of the estimated coefficients.

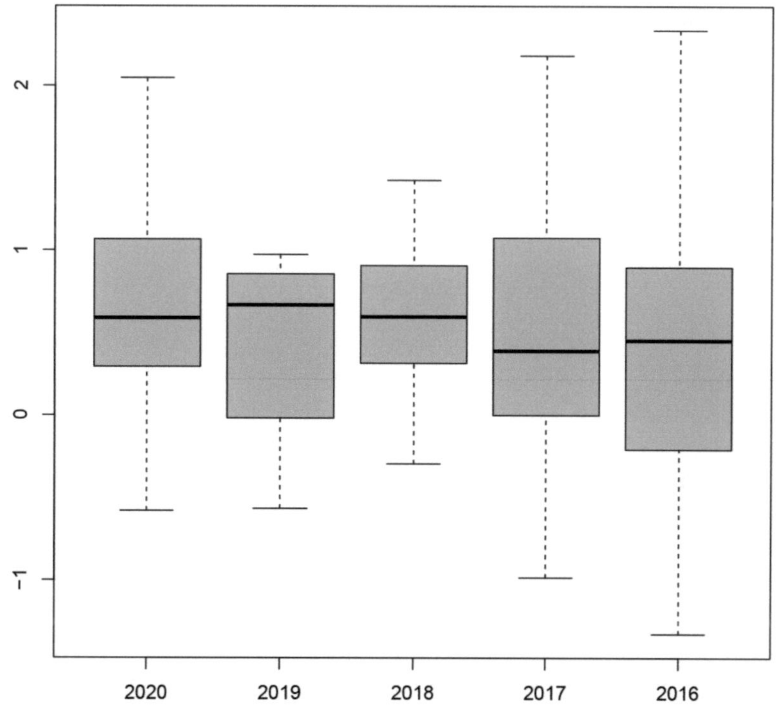

Figure 4.
Box plot of the intercepts υ^*.

	2020	2019	2018	2017	2016
μ^*	0.144	0.166	0.15	0.119	0.11

Table 5.
Results of the estimated expected value of the logarithm of inefficiencies μ^*.

in **Table 4**. Since the coefficients characterize the tangent hyperplane of the concave SMDDF of each DMU, DMUs with extreme values of the estimated coefficients may affect the shape of the estimated concave SMDDF and be identified as efficient DMUs. Indeed, all DMUs with extreme estimated coefficients are also evaluated as efficient DMUs in the deterministic scenario, as shown in **Table 2**. Moreover, we found that DMU02 is efficient in the deterministic scenario for the period 2018–2020. This observation is consistent with the results of the stochastic scenario. From **Table 4**, the period when DMU02 had extreme values of the estimated coefficients was also 2018–2020. However, the same is not true for DMU11. DMU11 was efficient in the deterministic scenario in 2016, 2017, and 2020, whereas its estimated coefficients were extreme only in 2020. However, this does not mean that DMU11 is not efficient, because technically, we cannot exclude the possibility that DMUs without extreme estimated coefficients cannot affect the shape of the estimated concave SMDDF. Regarding the efficient DMUs without extreme estimated coefficients in **Table 2**, they may be overestimated because the deterministic model ignored the impact of random errors.

The results of the deterministic scenario show that the Japanese life insurance industry exhibited worse performance between 2016 and 2020. By contrast, the results of the stochastic scenario in **Table 5** show that the estimated average inefficiency μ^* increased from 2016 to 2019 and then decreased, which implies that the Japanese life insurance industry likely performed better in 2020 than in 2019. Since the efficiency scores in 2020 exhibited slightly more variation than those in 2019, as shown in **Table 2**, the difference in the average performance between the deterministic and stochastic scenarios may be due to the impact of random error. Nevertheless, considering that both approaches have a small range of variation in the average (in) efficiencies (i.e., the deterministic scenario: $[0.004, 0.08]$; the stochastic scenario: $[0.009, 0.056]$), we conclude that the Japanese insurance industry exhibited relatively stable performance during 2016–2020.

5. Conclusions

This study considers a flexible structure of production technology that is constructed based on geometric convexity assumption. We investigated the relationship between the generalized multiplicative directional distance function (GMDDF) and geometric distance function (GDF) based on a given geometric convex production possibility set. We show that, given a deterministic scenario, the multiplication of GMDDF and GDF for the observed DMUs has the same value as the GMDDF calculated for the projection points. Based on this relationship, the GDF efficiencies can be easily measured by a DEA methodology. We also proposed a stochastic multiplicative directional distance function (SMDDF) in which all inputs and outputs are considered random variables. Given a stochastic scenario, we show that the SMDDF can be formulated as a CNLS problem. The productive performance of Japanese life insurance companies was investigated. Several companies were identified as efficient DMUs from both the results of the deterministic and stochastic scenarios. These companies can be used as projections for improving inefficiencies. Moreover, it seems that Japanese life insurance companies have a good performance in funding-related inputs since “total shareholders’ equity” and “operating expenses” were more efficient than were other inputs and outputs. By contrast, many of the Japanese life insurance companies may have suffered from poor performance in human resources. Furthermore, both the deterministic and stochastic scenarios suggest a decline in the average performance of the whole life insurance industry during 2016-2019. However, since the variations of average (in)efficiencies over the years are small, it is reasonable to conclude that the Japanese life insurance industry achieved relatively stable performance in the previous five years.

Abbreviations

CNLS	convex nonparametric least squares
DEA	data envelopment analysis
DMU	decision-making unit
FSA	financial services agency
GDF	geometric distance function
GMDDF	generalized multiplicative directional distance function
LIAJ	Life Insurance Association of Japan
NLP	nonlinear programming
QP	quadratic programming
SMDDF	stochastic multiplicative directional distance function
StoNED	stochastic nonparametric envelopment of data

Author details

Yu Zhao
Tokyo University of Science, Tokyo, Japan

*Address all correspondence to: yu.zhao@rs.tus.ac.jp

References

[1] Mehdiloozad M, Sahoo BK, Roshdi I. A generalized multiplicative directional distance function for efficiency measurement in DEA. European Journal of Operational Research. 2014;**232**(3): 679-688

[2] Portela MCAS, Thanassoulis E. Profitability of a sample of Portuguese bank branches and its decomposition into technical and allocative components. European Journal of Operational Research. 2005;**162**(3): 850-866

[3] Charnes A, Cooper WW, Rhodes E. Measuring the efficiency of decision making units. European Journal of Operational Research. 1978;**2**(6): 429-444

[4] Banker RD, Maindiratta A. Piecewise loglinear estimation of efficient production surfaces. Management Science. 1986;**32**(1):126-135

[5] Aigner D, Lovell CK, Schmidt P. Formulation and estimation of stochastic frontier production function models. Journal of Econometrics. 1977; **6**(1):21-37

[6] Kuosmanen T. Representation theorem for convex nonparametric least squares. The Econometrics Journal. 2008;**11**(2):308-325

[7] Banker RD, Charnes A, Cooper WW. Some models for estimating technical and scale inefficiencies in data envelopment analysis. Management Science. 1984;**30**(9):1078-1092

[8] Färe R, Grosskopf S, Lovell CK. The Measurement of efficiency of production. Vol. 6. Dordrecht: Springer Science & Business Media; 1985

[9] Majumdar M, Roy S. Non-convexity, discounting and infinite horizon optimization. Journal of Nonlinear and Convex Analysis. 2009;**10**:1-18

[10] Valadkhani A, Roshdi I, Smyth R. A multiplicative environmental DEA approach to measure efficiency changes in the world's major polluters. Energy Economics. 2016;**54**:363-375

[11] Roshdi I, Hasannasab M, Margaritis D, Rouse P. Generalised weak disposability and efficiency measurement in environmental technologies. European Journal of Operational Research. 2018;**266**(3): 1000-1012

[12] Afriat SN. Efficiency estimation of production functions. International Economic Review. 1972;**13**(3):568-598

[13] Kuosmanen T, Johnson A. Modeling joint production of multiple outputs in StoNED: Directional distance function approach. European Journal of Operational Research. 2017;**262**(2): 792-801

[14] Fukuyama H. Investigating productive efficiency and productivity changes of Japanese life insurance companies. Pacific-Basin Finance Journal. 1997;**5**(4):481-509

[15] Nobuyasu U. Japan's insurance market 2021. Tokyo: The Toa Reinsurance Company, Limited; 2021

[16] Eling M, Luhnen M. Frontier efficiency methodologies to measure performance in the insurance industry: Overview, systematization, and recent developments. The Geneva Papers on Risk and Insurance-Issues and Practice. 2010;**35**(2):217-265

[17] Portela MCAS, Thanassoulis E. Malmquist indexes using a geometric distance function (GDF). Application to a sample of Portuguese bank branches. Journal of Productivity Analysis. 2006; **25**(1):25-41

Chapter 6

Managing Large-scale Societal Change

Yiannis Laouris

Abstract

In this chapter, we discuss three decisive parameters for successful large-scale societal interventions: 1. The selection of the most representative and relevant stakeholders; 2. The application of an appropriate systemic problem-structuring methodology; and 3. The process used to convert the results of a structured deliberation into a clear strategy accompanied by a roadmap consisting of the most effective actions. We claim that the structured democratic dialog process emerges as an excellent tool for managing diverse types of societal interventions. Two models of intervention for large-scale societal reforms are briefly presented and discussed. The first is based on a quasi-synchronous process using the same intervention delivered at multiple localities. The second starts with one intense focal intervention and a process design that allows it to replicate and expand by creating spin-off agents or communities of change. The chapter concludes with recommendations.

Keywords: societal reforms, structured democratic dialog, problem structuring methods, community operations research, collective intelligence, collective wisdom

1. Introduction

As our world is constantly evolving, it will, of course, always be in need of changes and reforms. However, the increasing rate of change in combination with the rapidly increasing complexity renders the need for effective large-scale reforms a pressing emergency [1–3]. International bodies such as the UN, the EU, the G7/20, etc., as well as scientists, philosophers, and activists, alert us to the need for positive change in virtually all aspects of our lives. Reforms are especially critical in (global- and local scale) governance, peace, education, economy, and health. However, we still lack solid scientific models on how to plan and successfully implement large-scale reforms. Also, the number of successful and scientifically validated large-scale interventions is still small. On the other hand, the advances in technology and especially social media render changes previously thought as unachievable entirely feasible. The question is, however, whether we, as a society, must passively welcome changes just because of their feasibility? Does the fact that they are possible mean that they are also desirable or inevitable? Although not the focus of this chapter, humanity is at a point from which it could design and construct positive futures in the realm of what we call conscious evolution [4, 5]. To be able to achieve this, we need visionary leaders but, more importantly, appropriate science, methods, and tools. Effective leaders engaging in large-scale reforms rely on the triad, *will*, *ideas*, and *execution* for guidance [6, 7]. Clear, quantifiable, and ambitious vision and goals are crucial to building will. Those leading should express confidence in their people's creative potential and goodwill and know when and how to celebrate success. They should put systems and procedures in place to empower those whose lives will be influenced by any reforms to offer their ideas without fear of criticism. They should also allow them to hear or experience others' innovations and success stories. The execution, vis-à-vis, implementation, requires strategy, roadmaps, continuous attendance to processes, and willingness to adapt to new situations. Finally, the encouragement of cooperation and dialog among groups pursuing similar or identical goals increases the speed and quality of the change process.

In practice, there are many approaches to implementing a desired change. The first is by executive order coming from a gifted leader or from the management/government. Such interventions do not need the "acceptance" of the public, nor do they require lengthy approval processes. They are typical in hierarchical organizations and in emergency situations, such as the recent COVID-19 pandemic, and are often temporary or serve a short-term purpose. Sometimes, they are unwelcome and might even create societal unrest.

The most typical scenarios of societal change are through some form of policies or legislation, which could take place with or without the participation of all relevant stakeholders. In the latter case, the path can be rough (see next section). In this chapter, we are concerned with the case when relevant stakeholders are invited to define the problem collectively but also deliberate to come up with a consensus as to what actions are required to move forward. Thus, the first challenge is the selection of the most *representative* and *relevant* stakeholders.

The second challenge for achieving positive change is the availability and application of an *appropriate systemic problem-structuring methodology* and *relevant tools*. And the third and probably greatest challenge is how to convert the results of any process of deliberation into a clear strategy accompanied by a roadmap consisting of the most effective actions. We claim that the structured democratic dialog (SDD) process emerges as an excellent tool for addressing these three challenges and successfully managing diverse types of societal interventions.

In this chapter, we present and discuss two models of intervention for large-scale societal reforms. The first is based on a quasi-synchronous process using the same intervention delivered at multiple places. The second starts with one intense focal intervention and a process design that allows it to replicate and expand by creating spin-off agents or communities of change. The chapter concludes with recommendations.

1.1 Stakeholder participation: authenticity, diversity, and equity

Researchers [8] have identified factors that impact large-scale change already in the 90s and they mostly point to human factors. The lack of support by the top management, their attempt to force change, inconsistent actions, unrealistic expectations, absence of meaningful participation, poor communication, unclear purpose, and misplacement of responsibility were found to have a highly negative impact. Factors with a highly positive impact include the management's tangible and visible support and commitment, good preparation, encouragement of stakeholder participation, a high degree of communication, a reward system, etc. Methods for stakeholder identification and engagement have been well described by Gregory et al. [9]. All of the above boil down to what we call *authentic participation*. Indeed, the law of requisite action [10] of the science of SDD[1] predicts that "action plans to redesign complex

[1] For a brief introduction to the first six laws see [23].

socio-technical systems without the authentic and true engagement of those whose futures will be influenced by the change are bound to fail." One can think of countless empirical examples.

In many cases reported in the literature, the "type" of stakeholders engaged was not made explicit, and their role was not specified [11]. Also, certain (possibly marginalized) groups might be excluded for justified or unjustified reasons [12]. A diversity of stakeholders is imperative if all perspectives are to be considered, in line with the law of requisite variety [10, 13–15].

In sum, the effective, democratic, and equal participation (i.e., equity criterion) of all those who have a stake has the great advantage that assuming consensus is reached, the great majority backs up the decisions for change. Consequently, the change has a higher chance of surviving for extended periods. Some ways by which the application of the SDD methodology guarantees the above are briefly explained in the next sections.

1.2 Choosing an effective problem structuring methodology

The management of complex social problems requires the application of a systemic problem-structuring method (PSM). Matching an appropriate PSM and tools to a specific system is of paramount importance [16]. Polls, surveys, interviews, focus groups, etc. might be appropriate to access the "general attitudes" of the stakeholders or the wider public. If, however, the aim is to enable them to exchange points of view, increase trust between conflicting groups, attempt an in-depth analysis of the issue or challenge at hand, deliberate, consider pros and cons of different solutions, and make choices or informed decisions, then participatory methods, such as world cafés, citizens assemblies, operation research and management science (OR/MS) methods, SDD, etc. become a must. In most cases, different methods and tools need to be used in different phases of the process. Operation research scientists [17] argue that OR/MS methods and tools are often used not so much because of their effectiveness but more for "prestige." They observe that there is an inverse relationship between the importance of social reforms and the use of OR/MS techniques for their management. Those in charge of change prefer to use political tools (which also serve their "masters") to avoid addressing the real conflicts, or dealing with the low degrees of bureaucratization. In this chapter, we argue that the structured democratic dialog process has emerged as an excellent PSM capable of managing any type of large-scale socio-technical reforms, exactly, because on the one hand, it relies on well-founded and repetitively validated tools and processes, and on the other hand, it is particularly effective in interconnecting different points of view and positionalities, thus resolving multiple conflicts of purpose and values and generating consensus [18, 24].

1.3 Converting strategies and roadmaps to tangible change

Probably the most significant challenge toward achieving an envisioned societal (or any other type of) positive future is that strategies and plans fail to produce the desired results. The problem lies in what we could call the conversion of a vision and associated strategy and individual steps of action toward achieving the desired goals into a coherent and effective sequence. The orderly (typically timewise) arrangement of sub-actions is called roadmaps, i.e., plans that articulate a specific course of action [19]. In our context, roadmaps could be orderly sequences consisting of goals, phases, processes, or milestones, collectively referred to as steps, each supporting the subsequent ones. A roadmap can also be defined as the compilation of views of a group of stakeholders as to what to do, when, who, and how to get where they want to go [20]. A plan must be supported and backed up by all stakeholders to be effective. In this paper, we argue that interpretive structural modeling (ISM) [21] in connection with the multi-parameter evaluation [22] of impact, feasibility, and probability of happening without intervention serves to help organize the most effective ordering of a roadmap's steps.

2. Models of intervention

We present two models of large-scale interventions. In the first (**Figure 1** left), the same type of intervention is delivered more or less synchronously to multiple groups that are distributed. The term distributed refers to either spatial (i.e., similar stakeholders but in different geographic locations) or contextual (i.e., stakeholders in different communities/sectors or with different perspectives or group interests). In the second model (**Figure 1** right), the process begins with one intense focal intervention and a process design that allows it to replicate and expand by creating spin-off agents or communities of change. The premise of both models is that the will, ideas, and pragmatics of implementing the change are generated (or facilitated vis-à-vis "implanted") and allowed to grow at multiple (spatial or contextual) localities. Awareness, perceptions, identified challenges, visions, and actions grow around each locality and assuming they are strong enough, they gradually connect to each other and eventually embrace the wider community of stakeholders. The approach resembles the metaphor of multiple forest fires (in our example in a positive sense) in which independent fires join forces to form an unstoppable fire. In the next sections, we present examples for both models from our own experience.

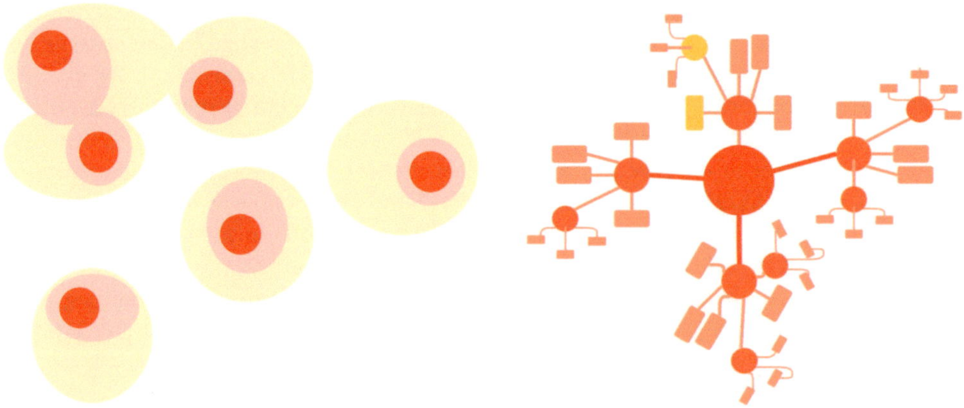

Figure 1.
Models of intervention for large-scale social changes.

3. The SDD methodology

The methodology and process of implementing either a single SDD or an array of parallel or sequential processes have been described elsewhere [14, 23]; for arrays of SDDs see [24, 25]. Here we provide a brief overview just sufficient to acquaint our audience with the key steps of a single SDD (**Figure 2**) or an array (for examples see **Figures 3** and **4**) of SDDs.

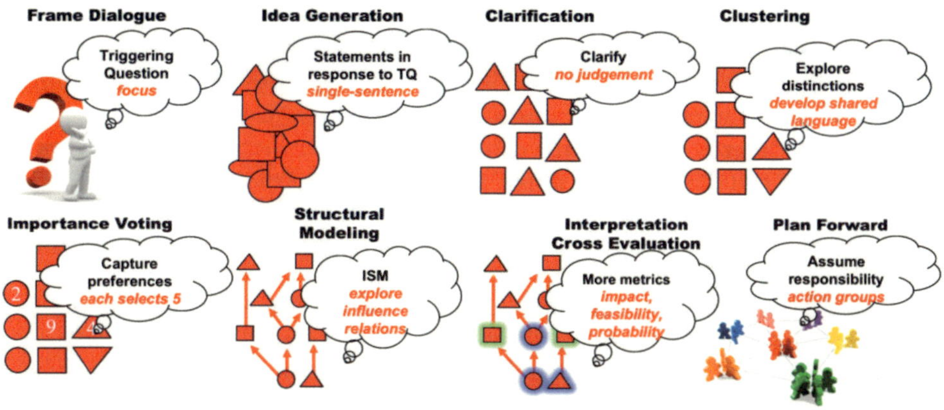

Figure 2.
Steps of a typical single SDDP implementation.

The first step of any SDD (**Figure 2**) is the generation of observations concerning the problematic situation in response to a triggering question (TQ). Each participant is invited in a round-robin manner to contribute only one response at a time in the form of a single statement, which should contain only one[2] specific observation. Contributions are numbered and their authors are registered. Giving individuals space to generate their ideas without criticism from others helps to counter groupthink and clanthink [27]. This approach satisfies not only the equity criterion (see 1.1) but also facilitates active listening and learning.

In a second step each participant clarifies her idea. This can be conducted through synchronous (f2f or virtual) processes, or asynchronously using the IdeaPrism[3] (or equivalent) app to record short videos, or through a collaboratively developed cloud-based document where everyone can edit, comment, or ask for further clarifications. During the clarification process, others can ask questions about meaning, but no judgment is allowed. This facilitation technique is intended to protect the autonomy and authenticity of participants so that no participant is discouraged, and no idea is prematurely evaluated and/or rejected. The so-called law of requisite autonomy in distinction making guarantees that "during the dialogue, the autonomy and authenticity of each person contributing ideas are protected" [28].

The next step involves the categorization of observations using a bottom-up approach. This process takes much longer than top-down clustering methods because it encourages discussion. Evolutionary learning takes place as the participants are encouraged to explore how specific aspects of their ideas might make them similar to other ideas; a process that forces them to draw further distinctions. The Law of Requisite Evolution of Observations asserts that "The elemental observations made by stakeholders in the context of a complex design situation, are interdependent" [10, 14, 23]. During the above steps, the participants are invited to reconsider the importance they assign to various observations. While the authors are liberated to clarify their observations with even inventing their own language and collectively searching for similarities in their effort to create categories, they better understand each other's positions. Only after this,

[2] This is important because, when observations are examined for similarity between them or influence on one another, if one statement contains several ideas or is too general, the process is compromised.
[3] IdeaPrism is a free app available in app stores.

individual participants are requested to choose typically five out of the total set of ideas according to their perceived importance. Known as Boulding's [29; see also 8, 10, 14] law of requisite saliency, the law refers to the range of importance that people assign to observations relative to other observations. The relative importance of an idea can be understood only when it is compared with the ideas of others (it is rare for people to choose only their own ideas as most important). All the ideas that receive a threshold number of votes (i.e., those that participants consider the most important) enter the ISM process, which comes next. The threshold number of votes is determined as as a function of the amount of time available to conduct the ISM.

In ISM, participants are confronted with two ideas at a time (to reduce cognitive load in recognition of our human limitations; i.e., Miller's Law of Requisite Parsimony; [30]) and are requested to decide whether one influences the other. In a synchronous implementation, they are invited to present arguments pro or against a relationship and engage in discussion. In our newly developed asynchronous models [31] participants may conduct parts of this process individually.

A relation is established only when it is supported by a large majority (typically 75%) following the constructive deliberation. The application of Warfield's [21] ISM algorithm reduces the number of questions that the software will ask. The binary connections that are established by the group are used to build up an influence map (see the example in **Figure 3**). Meaning and wisdom are produced only when the participants begin to understand the relationships (such as similarity, priority, influence, etc.) among their different ideas. The influence map reflects the shared understanding and the consensus of the participants.

Since challenges at the bottom of the structure correspond to the root causes of the problem, the method is also referred to as "root cause mapping." When the SDD is about exploring actions, the factors at the root are referred to as "deep drivers." The factors that end up at the root of the map are the ones with the greatest influence.

Participants engage in further discussions on how to resolve the obstacles at the root, and as these influence all the problems further up the structure, the idea is that addressing the root causes should have positive knock-on effects throughout the interlinked system of issues that the participants want to tackle. Analogously, factors at the root of an actions' SDD should be given priority when developing action plans. This form of problem structuring helps to minimize the erroneous priorities (first observed by Kevin Dye; see [24]) effect, which comes into play when strategic actions are targeted at isolated aspects of the problematic situation without their interconnections being considered.

Our team has applied the SDD in more than 100 different contexts, including peace and conflict resolution [32–34]; government and societal challenges (e.g., "Wine Villages" [25] and "Merging of taxation systems," conducted by CAPA [35]); discovering and collectively agreeing on research agenda priorities, thus influencing European Commission funding [36]; the support and capacity building of youth and civil society [37]; Uniting for Citizenship and Participation [38]; envisioning and designing new educational systems (Reinventing Education, [39]); and reinventing democracy [18, 26, 40]. For a complete list of Future Worlds SDDP applications, see Footnote.[4]

[4] https://www.futureworlds.eu/wiki/Chronological_List_of_SDDPs_by_Future_Worlds_Center_and_Associates

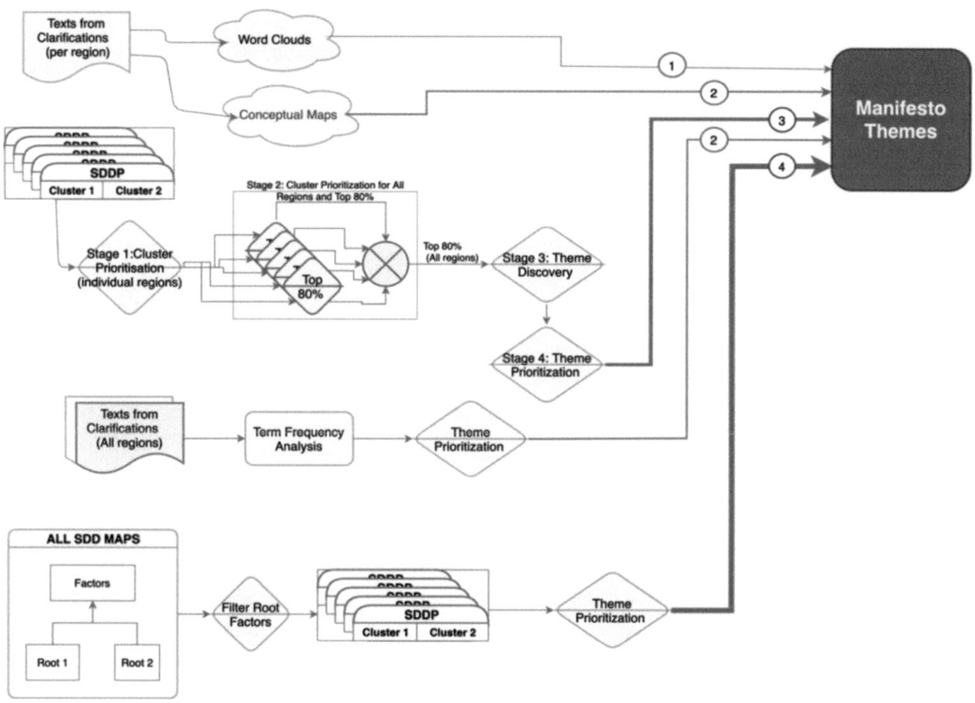

Figure 3.
Reprinted from reinvent democracy manifesto [26].

4. Real-world examples of large-scale interventions

In the following sections, we briefly introduce two examples, one international and one national, for each type of intervention.

4.1 Quasi-synchronous, spatially distributed interventions

4.1.1 Reinventing democracy bottom-up

The "Reinventing Democracy in the Digital Era" project [18, 26, 40] was a global project funded by the UN Democracy Fund. The aim was to identify shortcomings of our current systems of governance that could be improved through technology, and come up with concrete actions, policies, etc. capable of alleviating them. The process engaged about 1000 youth, from about 50 countries, in structured, face-to-face as well as virtual deliberations (**Figure 3**).

Selection of participants: Eleven weighted criteria were applied to the evaluation of a few hundred applicants in order to choose ca 20 Core Participants per region (i.e., five regions: Africa, Europe, Latin America, MENA, and Australasian; a total of 100). Gender (weight .2) and age (.15; young people 18–30 years old) had the largest weights. Four (i.e., anti-discrimination criteria; years of relevant experience or/and prior relevant activities; potential for organizing follow-up activities; availability of sponsors) had a weight of .1, and the remaining five (i.e., belonging to associations with wide networks; communication skills; reliability/commitment; country of origin/nationality; uninterrupted access to social networking) were weighted with .05.

Each core participant had to use similar criteria to invite 10 so-called shadow participants (who contributed throughout the 2-year period virtually), thus bringing their number to ca. 1000.

Methodology: Five identical interventions have taken place in the five different regions over a period of two years with the core participants. Week-long sessions followed the same format: Two days were spent on a critical systemic examination of shortcomings, with a view to considering the potential for improving democracy in the digital age. Then two days were spent on developing a collective understanding of the "deep drivers for change," which could serve as an inspiration for significant action to be pursued (by the participants or by others inspired by the maps), which were later made accessible on the internet via a Manifesto [26]. On the fifth day, participants formed groups to create action plans for themselves to pursue, springing from the collective work in locating leverage points for significant types of action while working on the second TQ. The last day was reserved for exploration of the city and joint activities to facilitate bonding.

Action: To compile the hundreds of shortcomings and actions (from a total of 10 SDDs) proposed by the participants into a short and clear actionable whole, we applied a novel multi-methodology for data compression (**Figure 3**). Five, weighted, data processing methods (the numbers on the arrows pointing toward the "Manifesto Themes" box in **Figure 3** correspond to the weights) were applied to extract seven prevailing themes, which were summarized in the Manifesto [26], where also the details of the methodological process are explained.

4.1.2 Reforming local authorities in Cyprus

The objective of the "Provision of Services for the Diagnosis of Learning and Development Needs for the Local Authorities of Cyprus" project was to improve the skills of the human resources of the Cypriot Local Government Authorities (LGAs). The overarching goal was to strengthen their administrative and leadership capacity with the view of facilitating the clustering of services and the eventual merging of LGAs to optimize the services offered to the public. The project was implemented between 2009 and 2015 under the special objective "enhancing administrative abilities in services provided in the public sector." The Cyprus Ministry of Finance supported this initiative using a palette of European funding, including structural and social cohesion funds.

Selection of stakeholders: The participants were selected from local communities. They included mayors, local authority employees, local educators, local and global business people and developers, local agricultural and other producers, repatriated citizens, older people, and youth. To secure requisite variety of their conflicting interests and objectives, particular emphasis was placed on conflicts between central and local governments, conflicts between neighboring local authorities, and conflicts between different stakeholders within the same local community.

Methodology: In this model, the diagnosis of the learning and development needs of the LGAs was embedded within a rich set of activities (**Figure 4**). It was preceded by identifying best practices in the EU member states and face-to-face interviews with LGA officials (left boxes in **Figure 4**). These actions informed the selection of the most relevant stakeholders and supported the framing of the TQ.

Action: Following the implementation of 10 SDDPs across the country, a network was set up comprising regional learning management groups and a coordinator for each group to support the management of learning in LGAs (see right side of **Figure 4**). Learning

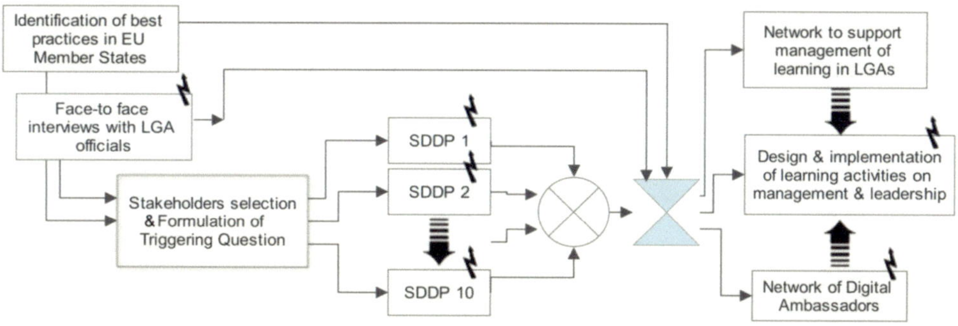

Figure 4.
Reprinted from [24].

activities were designed and implemented to enhance management and leadership. The overall intervention was supported by a strong awareness campaign, the engagement of political and government agencies, the expertise of technical advisors, and a network of digital ambassadors, many of them visiting foreign countries to research best practices. The intervention has eventually facilitated political decisions (in 2021/2022) for reforming the local authorities [24].

4.2 Single-focal intervention followed by evolutionary replication

4.2.1 Facilitating peace dialog across war zones

The "Civil Society Acts Beyond Borders" project [34] was funded by the European Instrument for Democracy and Human Rights of the European Commission. The aim was to empower civil society actors, youth, and local authorities in Israel and Palestine to actively promote human rights and democratization.

Selection of stakeholders: Participants were selected from diverse local communities by expert informants. The key criterion was to secure a mixture of peace activists but also a few who were against any peace "rapprochement."

Methodology: The project was designed as participatory action research grounded on the science of dialogic design. The project began with three consecutive (within a week) SDD processes (develop shared vision, identify obstacles, and explore actions) outside the conflict zone (in this case, Cyprus) engaging committed, already active peacebuilders (see left side of **Figure 5**). Between the second and the third SDD, the participants co-organized an international peace conference together with Greek and Turkish Cypriot actors. Having identified four target groups (business, informal education, students, and women) they created action groups (i.e., AG labels in **Figure 5**).

Action: Each AG selected and invited additional participants and organized their own SDD process, which was then followed by a comprehensive set of participatory intra-communal nonformal workshops (NFWS boxes in **Figure 5**). In parallel, a number of other activities were taking place. Following a training of trainers (top of **Figure 5**), those trained implemented a number of (up to 10) civil society strengthening pieces of training (purple boxes with a "W" in **Figure 5**), as well as multi-communal public debates (blue boxes with a "PD" in **Figure 5**), and a conference (purple box in **Figure 5**).

The above design offered the conditions for the objectives of the project to go viral, engaging several thousand people in various activities, i.e., to continue through self-replicating processes beyond the lifetime of the project.

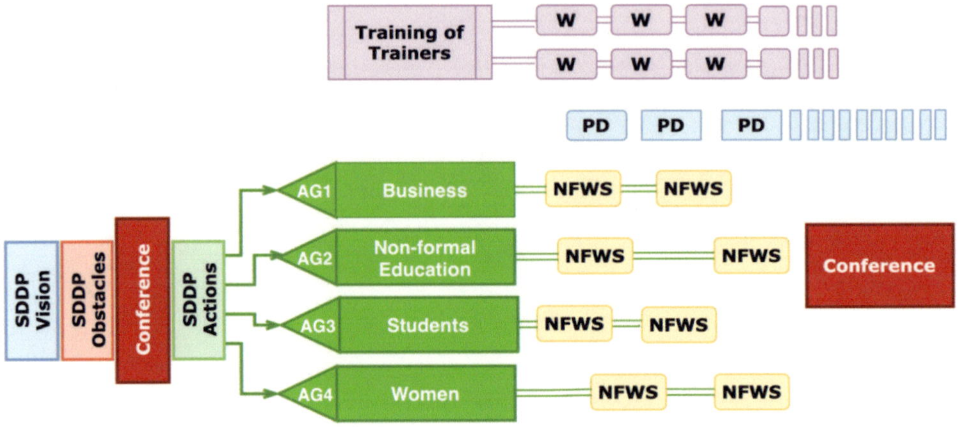

Figure 5.
Organization of activities in the CSABB project (from [34]).

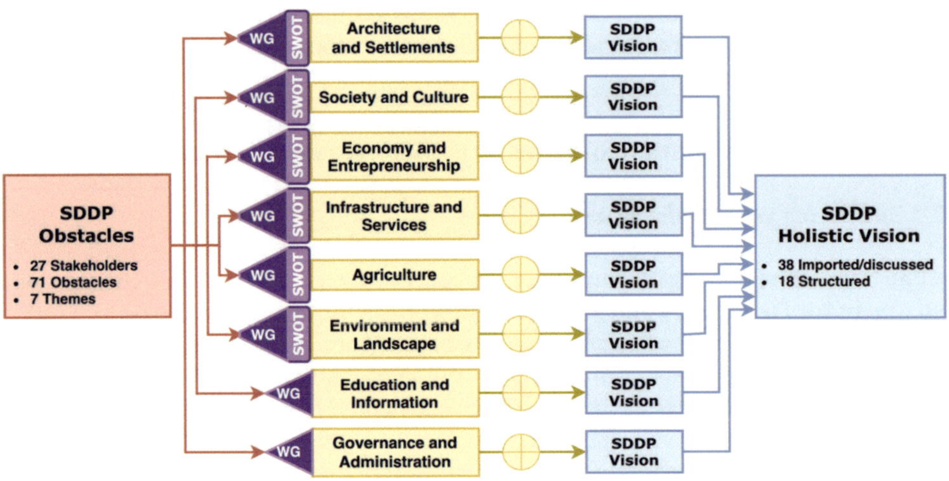

Figure 6.
Organization of activities in the wine villages project.

4.2.2 Local development of the wine villages in Limassol, Cyprus

The "Limassol Wine Villages Local Development Pilot Project: the contribution of heritage to local and regional development" was implemented through funding given by the Council of Europe's Local Development Pilot Projects Program[5] to the Department of Town Planning and Housing, Ministry of Interior of the Republic of Cyprus.

Selection of stakeholders: This project was implemented in phases. For the first SDD, 29 participants, representing many stakeholders' organizations were selected following standard stakeholder identification techniques [9]. In the following phases, a total of more than 150 stakeholders 'entered the scene' when they were acknowledged as stakeholders by other stakeholders [41].

Methodology: The intervention began with an SDD using TQ: "Which are the obstacles to the development of the Wine-villages of Limassol?" The participants

[5] https://www.coe.int/en/web/culture-and-heritage/ldpp

identified 71 obstacles (**Figure 6**; left), which they clustered into 7 themes of sectoral (thematic) or strategic nature (society and culture; architecture and settlements; economy and entrepreneurship; infrastructure and services; agriculture, environment, and landscape; education and information; and governance and Administration; second column in **Figure 6**). In order to develop each theme, a respective working group was created, which engaged further stakeholders thus allowing the base to drastically enlarge. Most working groups conducted a SWOT analysis (to obtain a better understanding of the area's real limitations and potentials at the local and macro-regional level) followed by a Vision-SDD using as TQ "What are the descriptors of the desired situation for the sustainable development of the wine villages in the field of... (relevant thematic)? " aiming to develop a shared vision. In a concluding SDD, the most influential factors from each Thematic SDD were selected and structured to develop an overall vision across all thematic areas (**Figure 6**, right).

Action: In total 150+ individuals from ca 75 different stakeholders' groups were involved in 10 SDDPs and 5 SWOT analysis workshops. The intervention allowed stakeholders to get involved in the policy process and in the formulation of the vision along with the strategy and its objectives. The model promoted territorial sensitivity and fostered identity and heritage. Most importantly, the intervention began with only 29 actors and ended up engaging actively more than 150.

5. Discussion

A large-scale societal intervention aims to support a community of stakeholders to develop a shared understanding of their problematic situation, converge on a clear shared vision, and ultimately generate collaborative action toward a desired future. This chapter proposed two models for large-scale interventions and presented examples of applications. Both models replace the hierarchical management of change in the design arena (i.e., relying predominantly on leaders or experts) in favor of models that include all parties with a stake (i.e., the stakeholders) in the definition and resolution of complex issues confronting them. Our first model applies the same intervention delivered at multiple places quasi synchronously. The term "places" could denote geographical or communal localities, but it could also denote corporate, religious, secular, local or international groups, etc. Our second model begins with one intense focal intervention using SDD and a design that allows the momentum generated to replicate, migrate, and/or expand by creating spin-off agents or communities of change. Assuming the momentum is strong enough, the waves of change will gradually reach each other and merge to embrace the wider community of stakeholders.

Both models presented here have not achieved extensive social change. The first reason is that the number of stakeholders engaged was a few hundred; i.e., too small compared to the total population. Second, some interventions were not extensive enough to generate a robust social wave toward change, because they lack the necessary political support and will. Nevertheless, eventually reforms did take place, even with a delay. To accelerate the process and achieve positive social change in a fraction of the time, we should probably utilize more social media and virtual-hybrid applications. We also need a new theoretical grounding of massive collaboration. Challenges of scalability and applications that made the participation of large numbers of participants possible are discussed only scarcely (e.g., [15, 42]) in the literature. A recent example of a large-scale project combined SDD with an interactive software called "Pathways to Wellbeing," [43] where the latter was aimed at facilitating citizen involvement in thinking together about choices being made to move toward inclusive well-being. In other examples utilizing virtual communications, the author's team has experimented with organizing SDDs in Second Life[6] and/or engaging participants in asynchronous processes using asynchronous ISM and IdeaPrism [31]. Such attempts may violate to some degree Laws of SDD. For a critical review see [31]. An additional bottleneck is trust. Of course, when the envisioned change is constrained within a local environment, the issue of trust might be lessened because the participants might already know each other [44].

6. Conclusions and recommendations

Contemporary liberalism has progressed to the point that it inhibits collective decision-making that serves the common good. It also undermines collective responsibility. Everyone agrees that we can be free and diverse only to the extent that our freedom and diversity do not undermine the rights of others. We, however, lack models of governance capable of balancing individual vs. collective needs and interests. This chapter proposes that the SDD methodology, embedded within a larger framework of OR methodologies, is ideal for engaging large groups of stakeholders in productive and efficient dialogs for the collective and systemic definition, as well as the resolution of their issues. It ensures that the deliberation produces high-quality observations by imposing structure and discipline and not allowing them to converge prematurely to preconceived choices of issues, options, and solution alternatives. It also creates an environment that facilitates openness to the ideas of others, which is vital for next-generation democracy.

Agents of change can choose from a wide range of strategic designs for large-scale improvement, considering available resources and constraints. Examples range from executive mandates, which may be appropriate for specific, small-scale changes that can be immediately implemented in a hierarchical system, to campaigns, which may be applicable for medium-scale interventions that rely on broad will-building [45, 46], to large-case change, which requires bringing together teams of stakeholders from numerous, often interdependent sectors, for structured interactions and learning [47]. The SDD approach addresses complex challenges, including sustainability, climate change, democratic deficiencies, pandemics, etc., all of which require new forms of stakeholders' engagement to work across conceptual and spatial boundaries [48, 49]. Some of the examples presented here have failed to reach the threshold for achieving extensive social change, primarily not because of theoretical limitations but because they were not supported by those who have the power. Democracy will rise to the level of collective wise decision-making that serves the common good only when the people manage to hold their leaders accountable for their choices and actions and put novel systemic approaches, instruments, and tools in all democratic processes.

Acknowledgements

The implementation of the projects has been the result of a team effort of dozens of experts and associates acknowledged in cited project pages and/or publications.

[6] Second Life is a virtual 3D environment that allows participants choose their own avatars and work collaborative in virtual spaces that resample conference rooms.

The specific case studies received funding as follows: Reinventing Democracy in the Digital Era was funded by the UN Democracy Fund (Contract number: UDF-GLO-13-532), "Provision of Services for the Diagnosis of Learning and Development Needs for the Local Authorities of the Cyprus" was funded by the Republic of Cyprus (Contract number 6/2009), "Human Rights and Reconciliation – Civil Society Acts Beyond Borders" was funded by European Commission, European Instrument for Democracy and Human Rights (Contract Number EIDHR/2009/167-502), and "Limassol Wine Villages Local Development Pilot Project: the contribution of heritage to local and regional development" was funded through structural funds managed by Marios Michaelides, acting head of the Cyprus Academy of Public Administration and various Ministries.

Author details

Yiannis Laouris
Future Worlds Center, Nicosia, Cyprus

*Address all correspondence to: laouris@futureworldscenter.org

References

[1] Fullan M. The return of large-scale reform. Journal of Educational Change. 2000;**1**(1):5-27

[2] McCannon CJ, Berwick DM, Massoud MR. The science of large-scale change in global health. Journal of the American Medical Association. 2007;**298**(16):1937-1939

[3] Christakis AN. A retrospective structural inquiry of the predicament of humankind. In: Rescuing the Enlightenment from Itself. Boston: Springer; 2006. pp. 93-122

[4] Harris LD, Wasilewski J. Indigeneity, an alternative worldview: Four R's (relationship, responsibility, reciprocity, redistribution) vs. two P's (power and profit). Sharing the journey towards conscious evolution. Systems Research and Behavioral Science: The Official Journal of the International Federation for Systems Research. 2004;**21**(5):489-503

[5] Marx HB. Conscious evolution: The next stage of human development. Systems Research and Behavioral Science: The Official Journal of the International Federation for Systems Research. 2003;**20**(4):359-370

[6] Nolan TW. Execution of Strategic Improvement Initiatives to Produce System- Level Results. Cambridge, MA: Institute for Healthcare Improvement; 2007

[7] Resar R. Will, ideas, and execution: Their role in reducing adverse medication events. The Journal of Pediatrics. 2005;**147**(6):727-728

[8] Covin TJ, Kilmann RH. Participant perceptions of positive and negative influences on large-scale change. Group & Organization Studies. 1990;**15**(2):233-248

[9] Gregory AJ, Atkins JP, Midgley G, Hodgson AM. Stakeholder identification and engagement in problem structuring interventions. European Journal of Operational Research. 2020;**283**(1): 321-340

[10] Laouris Y, Laouri R, Christakis A. Communication praxis for ethical accountability: The ethics of the tree of action: Dialogue and breaking down the wall in Cyprus. Systems Research and Behavioral Science: The Official Journal of the International Federation for Systems Research. 2008;**25**(2):331-348

[11] Degeling C, Carter SM, Rychetnik L. Which public and why deliberate?–a scoping review of public deliberation in public health and health policy research. Social Science & Medicine. 2015;**131**:114-121

[12] Caroline W. Lee Five Assumptions Academics Make about Public Deliberation, and Why They Deserve Rethinking

[13] Ashby WR. Requisite variety and its implications for the control of complex systems. In: Facets of Systems Science. Boston: Springer; 1991. pp. 405-417

[14] Christakis AN, Bausch KC. How people harness their collective wisdom and power to construct the future in co-laboratories of democracy. In: IAP. 2006

[15] Laouris Y, Dye K, Michaelides M, Christakis AN. Co-laboratories of democracy: Best choices for designing sustainable futures. In: Social Systems and Design. Tokyo: Springer; 2014. pp. 167-183

[16] Solomon S, Abelson J. Why and when should we use public deliberation? The Hastings Center Report. 2012;**42**(2):17

[17] Papoulias DB, Tsoukas H. Managing reforms on a large scale: What role for OR/MS? Journal of the Operational Research Society. 1994;**45**(9):977-986

[18] Laouris Y, Romm NR. Structured dialogical design as a problem structuring method illustrated in a Re-invent democracy project. European Journal of Operational Research. 2022;**301**(3):1072-1087

[19] Kappel TA. Perspectives on roadmaps: How organizations talk about the future. Journal of Product Innovation Management: An International Publication of the Product Development & Management Association. 2001;**18**(1):39-50

[20] Phaal R, Farrukh CJ, Probert DR. Technology roadmapping—A planning framework for evolution and revolution. Technological Forecasting and Social Change. 2004;**71**(1-2):5-26

[21] Warfield JN. Toward interpretation of complex structural models. IEEE Transactions on Systems, Man, and Cybernetics. 1974;**5**:405-417

[22] Linstone HA, Turoff M, editors. The Delphi Method. Reading, MA: Addison-Wesley; 1975

[23] Laouris Y. The ABCs of the science of structured dialogic design. International Journal of Applied Systemic Studies. 2012;**4**(4):239-257

[24] Laouris Y, Michaelides M. Structured democratic dialogue: An application of a mathematical problem structuring method to facilitate reforms with local authorities in Cyprus. European Journal of Operational Research. 2018;**268**(3):918-931

[25] Wine Villages Cyprus; Series of SDDPs https://www.futureworlds.eu/wiki/Wine_Villages_Cyprus_Series_of_SDDPs

[26] Manifesto: Reinventing Democracy in the Digital Era. 2016. Available from: https://www.futureworlds.eu/wiki/Manifesto:_Democracy_in_the_Digital_Era. [Accessed: May 30, 2022]

[27] Warfield JN. Spreadthink: Explaining ineffective groups. Systems Research. 1995;**12**(1):5-14

[28] Tsivacou I. The rationality of distinctions and the emergence of power: A critical systems perspective of power in organizations. Systems Research and Behavioral Science: The Official Journal of the International Federation for Systems Research. 1997;**14**(1):21-34

[29] Boulding KE. The Impact of the Social Sciences. New Brunswick, NJ: Rutgers University Press; 1966

[30] Miller GA. The magical number seven, plus or minus two: Some limits on our capacity for processing information. Psychological Review. 1956;**63**(2):81

[31] Laouris Y. Method to integrate asynchronously produced individual influence maps into an extrapolated population influence map following the face-to-face stage of a structured democratic dialogue. Systems Research and Behavioral Science: The Official Journal of the International Federation for Systems Research. 2022

[32] Laouris Y, Michaelides M, Damdelen M, Laouri R, Beyatli D, Christakis A. A systemic evaluation of the state of affairs following the negative outcome of the referendum in Cyprus using the structured dialogic design process. Systemic Practice and Action Research. 2009;**22**(1):45-75

[33] Laouris Y, Erel A, Michaelides M, Damdelen M, Taraszow T, Dagli I, et al. Exploring options for enhancement of social dialogue between the Turkish and Greek communities in Cyprus using the structured dialogic design process. Systemic Practice and Action Research. 2009;**22**(5):361-381

[34] Civil Society Acts Beyond Borders. 2010. Available from: https: // www.futureworlds.eu/wiki/Act_Beyond_Borders. [Accessed: June 01, 2022]

[35] Merging of TAX - VAT Services in Cyprus. 2013. Available from: https:// www.futureworlds.eu/wiki/Merging_of_TAX_-_VAT_Services_in_Cyprus. [Accessed: June 01, 2022]

[36] CARDIAC - Advancing Research & Development in the area of accessible & Assistive ICT [Internet] 2010. Available from: https://www.futureworlds.eu / wiki/CARDIAC. [Accessed: June 01, 2022]

[37] Capacity and Synergy building among NGDOs and LAs in Greece, Cyprus and Malta for development [Internet] 2011. Available from: https:// www.futureworlds.eu/wiki/MeDevNet. [Accessed: June 01, 2022]

[38] UCYVROK - Uniting for Citizenship and Participation. 2008. Available from: https://www.futureworlds.eu/wiki/ UCYVROK__Uniting_for_Citizenship_and_Participation. [June 01, 2022]

[39] Reinventing Education. 2018. Available from: https://www. futureworlds.eu/wiki/Reinventing_Education. [Accessed: June 01, 2022]

[40] Reinventing Democracy in the Digital Era. 2016. Available from: https:// www.futureworlds.eu/wiki/Reinventing_Democracy_in_the_Digital_Era_(UNDEF). [Accessed: June 01, 2022]

[41] Goodman LA. Snowball sampling. The Annals of Mathematical Statistics. 1961;**1**:148-170

[42] Rosenhead J. Past, present and future of problem structuring methods. Journal of the Operational Research Society. 2006;**57**(7):759-765

[43] McIntyre-Mills JJ, Christakis AN. Social and environmental justice. In: From Polarisation to Multispecies Relationships. Singapore: Springer; 2021. pp. 283-307

[44] Ahlqvist T, Bäck A, Heinonen S, Halonen M. Road-mapping the societal transformation potential of social media. foresight. 2010

[45] Berwick DM, Calkins DR, McCannon CJ, Hackbarth AD. The 100 000 lives campaign: Setting a goal and a deadline for improving health care quality. Journal of the American Medical Association. 2006;**295**(3):324-327

[46] Wachter RM, Pronovost PJ. The 100,000 lives campaign: A scientific and policy review. The Joint Commission Journal on Quality and Patient Safety. 2006;**32**(11):621-627

[47] McCannon CJ, Berwick DM, Massoud MR. The science of large-scale change in global health. Journal of the American Medical Association. 2007;**298**(16):1937-1939

[48] Midgley G, editor. Systems Thinking. London: Sage; 2003

[49] Midgley G. Theoretical pluralism in systemic action research. Systemic Practice and Action Research. 2011;**24**:1-15

Chapter 7

The Impacts of the COVID-19 Pandemic in Supply Chain Operations

Alexsandro Silva Solon and Marcos Paulo Leite Silveira

Abstract

In December 2019, in Wuhan, China, the emergence of a new disease changed the lifestyle of the entire population. The COVID-19 pandemic has marked the world and continues to do so, causing several impacts in the most varied sectors. This chapter aims to show how the pandemic impacted the global logistics network by showcasing several supply chains that were affected, mainly focusing on e-commerce and food supply chains and their many aspects. This chapter also draws attention to how changes, caused by the virus, have transformed consumer behavior over the past 2 years. It will also present mitigation strategies that have been or could have been used by companies to minimize the risk of disruptions during the pandemic by creating a more resilient supply chain.

Keywords: supply chain, E-commerce, food supply chains, pandemic, resilience

1. Introduction

The year 2020 will remain in our memory forever. A pandemic which had never been seen in this day and age, took the whole world by surprise. It all started at the end of 2019, in Wuhan, China, where the first cases appeared. Thereafter, the virus spread around the globe, initially affecting Europe and then reaching Brazil in mid-March that very year. Within the peaks and valleys in the well-known contagion curve, and the emergence of variants which unleashed new waves of the virus, improvements can be seen in the national epidemiological scenario, thanks to the advance of vaccination. However, if we observe European countries, it is clear that there has been a significant increase in the number of people infected with the COVID-19 virus.

This whole scenario has triggered abrupt changes in our way of living, whether in people's relationships, in the way we work, attend classes, or do our shopping.

Everything has become digital and it is unimaginable to think that from now on, things will change or go back to the way they used to be.

Considering the general aspect of purchasing a product, the entire chain of events, from the beginning of the production process, up until delivery to the final consumer, many changes have taken place. Either, due to variations in terms of product demand, the increase of prices, or the lack of raw materials, all of which was inevitable, mostly because of the several factors involved in a production chain.

The purpose of this article is to analyze the impact of the pandemic on the Supply Chain, a term coined in the Unites States, which literally means "a network between a company and its suppliers to produce and distribute a specific product or service." Supply Chain refers to the set of production processes of a material, from the extraction of raw material to delivery of the final consumer.

The article will also analyze how certain aspects have been directly affected, more specifically, e-commerce, which is not only an important electronic sales channel, but also an opportunity to develop (Business to Consumer) B2C and (Business to Business) B2B companies. We also intend to look into how e-commerce was used as a reactive strategy to minimize the breakage risks during the supply chain cycles.

Furthermore, we intend to point out some changes in consumer behavior and how companies may adapt to the abrupt changes, in order to become resilient, in other words, to be able to reuptake their initial conditions, prior to the pandemic, and face these great transformations.

Subsequently, Section 2 will present the research methodology, while Section 3 will focus on literature review, followed by the analysis and discussions in Section 4. The final remarks and conclusions can be found in Section 5.

2. Methodology

This article aims to present how the COVID-19 pandemic has affected and continues to affect national and global supply chains, especially in the last link of the chain, the consumer. This section aims to show how this work was compiled.

An exploratory research was carried out to list, in amongst the countless existing supply chains, those which were directly impacted by the COVID-19 pandemic, such as a sudden increase in product demand, or several changes in consumer behavior during the most diverse stages of the pandemic. The research was conducted using a systematic search engine on targeted websites and applications, in which, by means of delimited keywords, it was possible to identify articles and reports published from January 2020 to October 2021, as specified in **Table 1**.

Among all articles found in the systematically structured research, it is possible to identify the effect of the COVID-19 pandemic on the supply chain in two specific sectors: e-commerce and food supply. In addition, as the aim was to globally assess these impacts, it is possible to collect data in the specified period from January 2020 to October 2021, simply by inserting specific keywords with names of countries, such as Brazil, United States, Mexico, England, Spain, France, Italy, for instance.

A bibliometric and exploratory research is of significant help in the decision-making process, because researchers can explore, organize and evaluate large masses of data which, if not evaluated with a more structured method, will not accurately generate such valuable results for decision-making [1]. By organizing this data, it is possible to reach a conclusion.

This work presents a combination of qualitative and quantitative research, as stated by [2], "qualitative research provides a better vision and understanding of the context regarding the issue, while quantitative research seeks to quantify the data by means of statistical analysis". Firstly, the whole pandemic scenario was contextualized, and afterwards, a data analysis was carried out on consumer behavior.

This study also intends to show, among a variety of risk mitigation strategies which happen during chain interruptions, those which successfully enabled

Sites and applications	Journals and academic magazines	Keywords
Scielo	International Journal of Production Research	Supply chain + COVID
Eric	International Journal of RF Technologies Research and Applications	Supply chain + e-commerce
Google Scholar	Emerald Insight: Discover Journals, Books & Case Studies	Supply chain + pandemic
Capes Brazil	International Journal of Supply and Operations Management	e-commerce + pandemic
BDTD	International Journal of Supply and Operations Resilience	Supply chain management + COVID
Research Gate	Brazilian Journal of Production Engineering	Supply chain management + COVID + impact
	Brazilian Journal of Operations & Production Management	Supply chain food + COVID

Source: The author.

Table 1.
Search structure.

companies to deal with disruptions and measures which could have been taken to avoid further losses and maximize profits.

Table 1 demonstrates the structure and research methodology applied, as well as sites, applications and keywords used.

3. Bibliographical review

This section will outline the literature review done, so as to present the main concepts present in the course of the work. The themes covered are as follows: supply chain, pandemic impacts in North America, pandemic impacts in Europe, e-commerce, pandemic impacts in e-commerce in Brazil and supply chain resilience (SCRes).

3.1 Supply chain

In [3] a supply chain is defined as "a network of organizations which are connected and interdependent of each other, working cooperatively to control, manage and improve the flow of materials and exchange of information from suppliers to end users".

A supply chain, in addition to all activities which it involves, such as input purchasing, storage, transport, packaging, sales and distribution, incudes all physical infrastructure and processes that support the operations. In other words, the supply chain is made up of physical elements, as well as planning and operational processes [4].

It is important to emphasize that a well-developed supply chain, which functions appropriately, is a major competitive differentiator. Lambert and Cooper, [5] claims that one of the most important paradigm shifts occur when companies, nowadays, do not compete with one another, nor does a product compete with another product, instead, the competition would be of one supply chain against another. That, in turn will reinforce the need to establish a strengthened chain management able to stand in good stead, should any adversities appear, as shown in **Figure 1**.

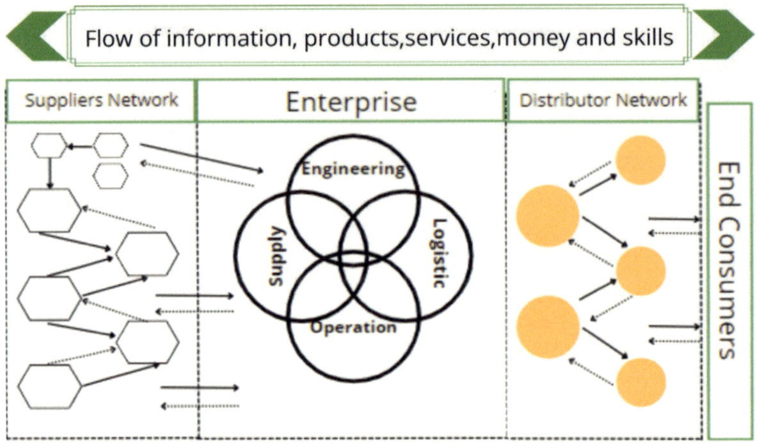

Figure 1.
Supply chain flow. Resource base: (Capacity, Information, "Core" competencies, Banking). Source: [6 , p. 9].

The several organizations involved, in addition to the large amount of information and materials that circulate both ways, only emphasize how dependent each member is of the other, and events which disrupt the balance of the chain can cause great harm to the parts involved.

3.2 Pandemic impacts in North America

According to [7], the COVID-19 pandemic has been one of the most significant events to disrupt the supply chain in recent history, and it is also likely to weaken many organizations globally. In an attempt to prevent the spreading of the virus, most governments around the world have responded with the implementation of human restriction measures. As there were no pharmacological treatments at the beginning of the pandemic, such as vaccines, border closures and quarantines, all of which proved to be extremely challenging to combat, as symptoms appeared at different levels of intensity in patients [7]. Global response has severely impacted international supply chain operations [7]. The National Association of Manufacturers (NAM) conducted a survey of its 558 members in the United States on the impact of COVID-19, and found that more than 78% of its members expected a serious financial impact on their businesses due to the uncertainty caused by the pandemic [8].

Another study by (PricewaterhouseCoopers) (PwC) reports that 87% of multi-sector companies in Mexico and the United States were very concerned about the disastrous impact of the pandemic [9].

As explained by Baldwin and Tomiura in a publication of 2020 [10], it can be predicted that the financial impact caused by COVID-19 on the manufacturing sector alone, will almost triple.

They state that, initially, the concentration of the disease was mainly in the world's manufacturing center (East Asia), and its spread to other industrial powers in the US and European Union (EU) would therefore create direct and massive disruptions in supply, which in fact did occur. Such sudden supply disruptions would then cascade to other industries in less affected countries due to the contagion effect on the supply chain. Eventually, macroeconomic declines in aggregate demand, along with delays in investment by companies, will undoubtedly generate countless disruptions.

Thus, both the vulnerability and the complexity of supply chains spread and triggered the effect of the pandemic, even in countries that initially did not suffer from the disease itself. China's large share of global trade and its expressive significance in the supply chain amplified the economic repercussions for other less affected markets. Even if the peak of the outbreak were short-lived (something that notably did not happen), with gradual recovery in production and demand in the coming months (which also did not occur), it still posed a substantial threat to global growth in 2020 [11].

No business sector was immune to the effect of the pandemic. Given the economic importance, considerable supply and demand shocks were not leveled in firms and industries. The impact of COVID-19 on different business sectors may vary due to differences in supply and demand patterns [12]. While some sectors saw a drop in demand, others saw a sudden increase. As people began to follow government calls to stay at home, some industries such as car manufacturing, crude oil production and transportation saw a decline in demand. In contrast, most tech companies experienced an increase in demand. Companies operating in the healthcare sector experienced a sudden surge in demand which has proved to be a tall order to meet in such a short period. Service companies that rely heavily on tourism and travel were also one of the first to suffer from a very strong and unexpected decrease in demand.

As this work unfolds, it aims to show specific cases where impacts of the pandemic are more evident, such as in e-commerce in Brazil, as well as certain effects of the pandemic in Europe because these were sectors where significant changes in consumer behavior were noticed, besides a substantial amount of other collected comparative data used to carry out the study. The response strategies used to mitigate the supply chain risks caused by the ongoing pandemic will also be respectively shown, compared and analyzed.

3.3 Pandemic impacts in Europe

A shortage in food is inevitable, considering the initial approach to combat the pandemic. The food supply chain is a complex web which includes producers, agricultural inputs, transportation, factories, distributors and retailers.

According to experts, restrictions on transport and other blocking measures, such as the implementation of greater control over sea cargo, the production process has suffered significant impacts. As a consequence, labor shortages started to unfold and prices peaked, which has led to food loss and waste. These obstructions normally prevent farmers from accessing markets, restricting their productive capacity and preventing them from selling their products [13].

As commuting and movement of people was restricted, restaurants that depended exclusively on face-to-face service were severely affected and had to be shut down. In Spain, for example, around 40,000 hotels, bars, cafés and restaurants had closed their doors by July 2020 (Le Parisien). To minimize this effect in England, the government gave a one-time cash donation of up to £25,000 to eligible restaurants, cafés, bars and pubs; a £1,000 bonus per employee to employers who retained employment contracts until 31st January 2021; besides paying 80% of salaries to employees placed on leave when establishments shut down because of the quarantine and other restrictive measures [14].

As a form of recovery, the UK instituted the "eat out to help out" project, in which the government paid up to 10 pounds for meals made from Monday to Wednesday in restaurants across the country [15].

Lockdowns reduced the movement of people and prevented some workers from reaching their workplaces. Many sectors, including the fuel industry saw a decrease in demand, as more people stay indoors, which consequently increased the price of fuels used by trucks to transport all sorts of materials.

On the other hand, the crisis also generated the need for innovations, creating opportunities for new business models in Europe and around the world [16]. Analyzed the implications of COVID-19 on food supply chains and reported that demand and supply shocks created during a pandemic mainly occur as a result of a change in consumer behavior. For example, the sudden urge to buy ready-to-eat foods caused several demand shocks, which led to labor shortages and disruptions to the transportation network. In addition, restrictions on transportation and delivery of goods across borders led to collateral shocks in food supply chains.

The change in consumer behavior can be seen in a survey carried out by the Reply platform, which collects data and performs analysis in the attempt to understand the progress of the pandemic and how it has affected society as a whole. Using the Quentin data research tool developed by TD Reply, which gathers data from Google Trends and Google Ads, it is possible to notice the effects. **Figure 2** shows how the COVID-19 pandemic impacted the Food and Beverage sector in Europe in 2021.

One of the first major changes that appeared at the beginning of the pandemic was how the population purchased food. As bakeries, markets and restaurants were closed, many people turned to the online format to buy their food. There was a 140% increase in Europeans' interest in buying food online in 2020 (data obtained through Quentin, a TD Reply's data research tool, Europe. 2019 x 2020).

The orange line represents the delivery rate of groceries, while the red line shows the delivery rate of ready-made meals. While the delivery of market products indicates

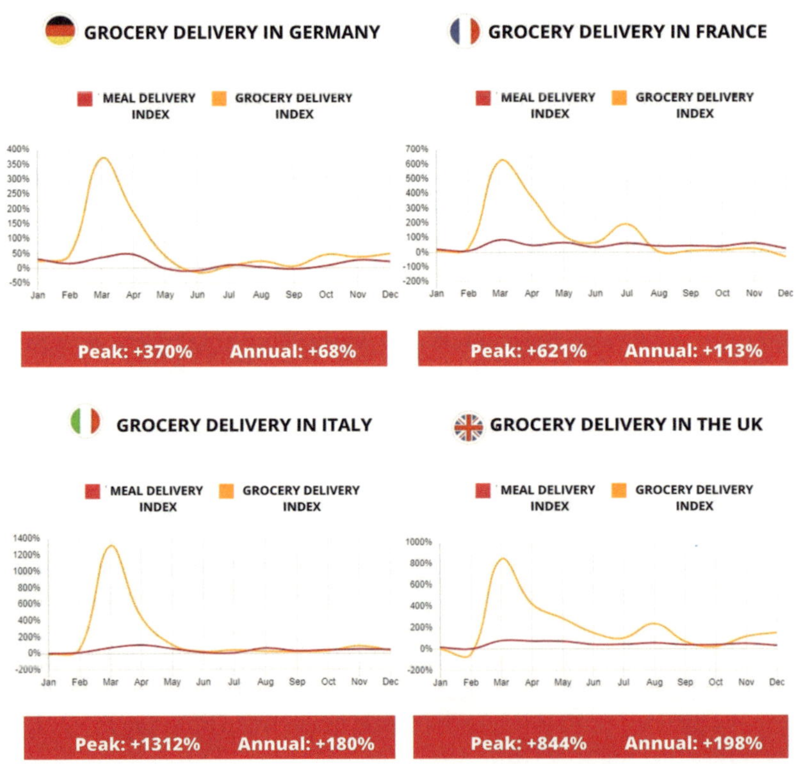

Figure 2.
Food delivery in Europe. Source: [17].

large peaks, especially at the beginning of the pandemic on the European continent, the delivery of meals did not suffer great variations throughout 2020. This indicated a possible tendency of the European people to experience food consumption in a different way, by ordering it from a specific restaurant and delivered to their homes.

Another aspect which underwent a major impact was how products were paid for. 87% of Europeans are willing to use less cash in the near future [18]. In addition, there has also been a greater concern with hygiene on behalf of establishments, in the attempt to reduce the circulation of banknotes, creating the perfect environment for the implementation and growth of new payment technologies, such as the QR code.

Figure 3 reveals how familiar the population was with the use of this technology in the same four European countries.

The orange line represents people who are not familiar with QR technology and the red line shows those who are, but do not know how to use it. Towards the end of 2020, those countries revealed a gradual drop in both lines, which indicates a possible increase in the population's knowledge on the subject.

3.4 e-Commerce

The e-commerce segment is an important sales channel and development opportunity for B2C and B2B companies.

It is impossible to imagine our current globalized world without the internet and all the transactions that take place in it. Besides the fact that the internet is undeniably helpful and offers a wide variety of benefits, the 'Internet of Things' (IOT) and data clouds are on the rise. Together they can integrate and improve communication and data storage of organizations, working hand-in-hand with each other. As the IOT is able to capture and process information, cloud technology transfers information to other devices.

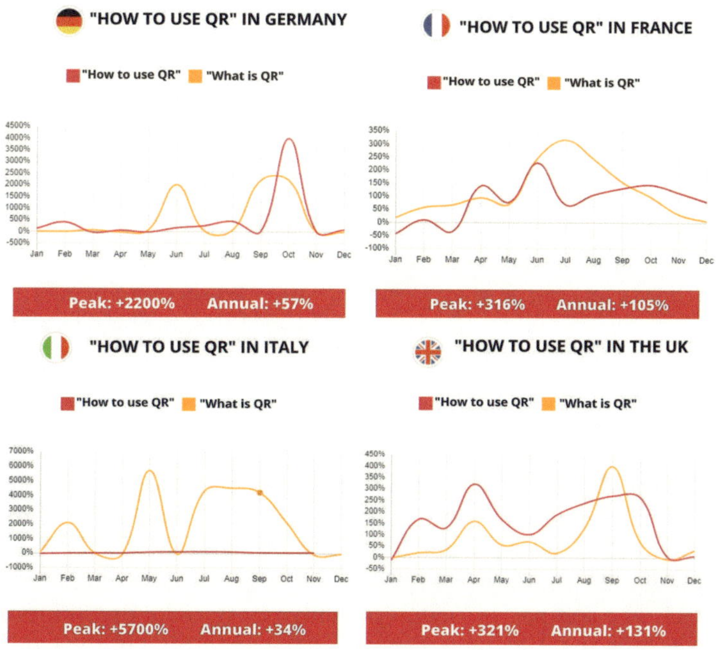

Figure 3.
Knowledge of the QR code in Europe. Source: [17].

The progressive digitization of commerce ensures fast communication between customers and companies and especially between companies that make up the same supply chain.

Organizations today aim to have a competitive edge over their contenders, continually considering the expansion of their businesses to reach different customers in different locations. This is not possible with a normal distribution channel, such as a physical store, or a single point of sale. This is achievable thanks to the Internet, available to anyone, anywhere. Similarly, modern supply chain management assists companies to reach an entire niche of customers in a healthy competitive way.

With an expressive rise in popularity in online sales, intensified mainly by the pandemic, further supply chain challenges have surfaced. Such challenges need to be resolved promptly in order for companies to be able to take advantage of the opportunities that e-commerce creates for them.

As the pressure to adapt and improve increases, e-commerce strategies create both difficulties and opportunities for companies that use this type of technology to boost sales.

Here are a few of the many types of relationships established between e-commerce practitioners [19]:

- C2C (customer to customer)—transactions between consumers on online sales platforms.
- C2B (customer to business)—encompasses transactions between consumers and companies, usually initiated by consumers to compare prices.
- C2G (client to government) –involves transactions between citizens and public administration, such as taxes.
- B2B (business to business)—implementation of business procedures between two companies.
- B2C (business-to-consumer or business-to-client)—classic business-to-consumer transactions.
- B2G (business to government)—transactions between businesses and public administration, such as public contracts or bids.

E-commerce was one of the most positively affected sectors during the pandemic. A remarkable increase in sales took place and the production chain was put to the test in order to meet the high variation in demand.

3.5 Pandemic impacts on e-commerce in Brazil

According to the Brazilian Chamber of Electronic Commerce (ABComm), more than 135,000 Brazilian stores adhered to e-commerce sales in the first three months of the pandemic in 2020, with the aim of remaining active even after their physical stores had been shut down. **Figure 4** represents the growth of e-commerce sales in Brazil since 2011. The comparison displayed only refers to the first semester. In other words, what we see is an immediate response to the beginning of the pandemic.

A report by *Setores do E-commerce no Brasil*, published and released by an agency named Conversion, in November 2020, proved that there was, in fact, a significant

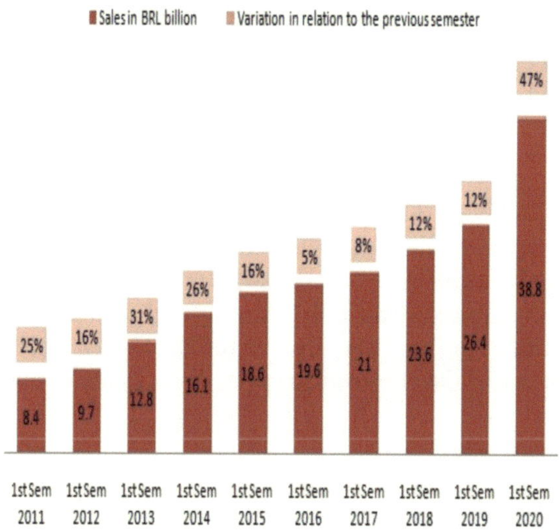

Figure 4.
The growth of e-commerce sales in Brazil. Source: Ebit | Nielsen via Meio & Mensagem.

growth during that period. Compared to the same period in 2019, 12 of the 15 sectors analyzed, presented a noticeable increase in online sales.

Pet products proved to be one of the markets with an expressive increase in e-commerce, with a leap of 60% in sales compared to 2019. The "food and beverage" and "home and furniture" sectors appear in second and third place, respectively.

After another year of COVID-19, with additional data collected from 2021 to be compared to, speculations that e-commerce would retract were utterly discarded. According to the report by *Setores do E-commerce no Brasil*, released by Conversion, a specialized agency in November 2021; e-commerce still presents positive rates although physical stores have been gradually reopening, with a 1.58% growth in October compared to September, that year.

Christmas, which is an extremely commercial period, can also be taken into account. **Figure 5** reveals how consumers preferred to do their Christmas shopping in 2020.

$I - 6 + n$ relation to the year 2021, a survey carried out by the National Confederation of Shopkeepers (CNDL) and the Credit Protection Service (CPS)

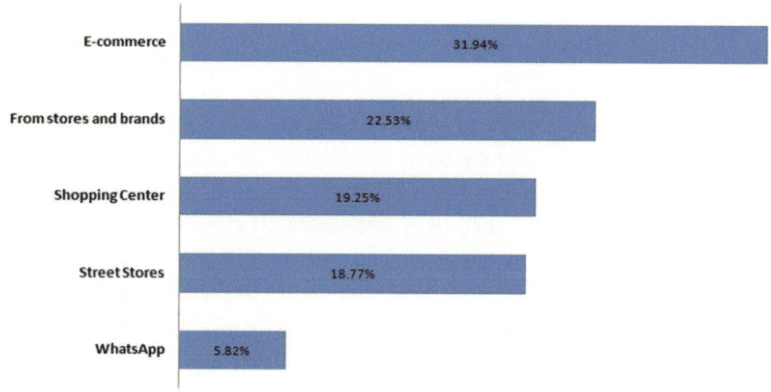

Figure 5.
Sales channels consumers intend to use for their Christmas shopping. Source: Conversion (2020).

reveals that 77% of Brazilians are likely to go shopping at Christmas. Following the trend of online shopping, (45%) of consumers will purchase their goods on the internet, followed by (43%) in department stores and (40%) in malls.

Among those who will shop online, 37% should purchase almost all presents on the internet; 31%, half their presents; and 21% are estimated to buy all their presents that way. On average, 67% of presents will be purchased online. With an expectation of moving BRL 64.8 billion in the economy, according to the National Confederation of Store Managers [20], with the vast majority of consumers still opting for online shopping, e-commerce is expected to maintain its strength this year. Even though vaccination rates have advanced and shopping in physical stores should be on the rise, the trend of increased online shopping, which started at the beginning of the pandemic, is far from being cast aside. On the contrary, it will continue to grow strong.

3.6 Supply chain resilience

In reference to [21] resilience is defined as: "The ability of a system to return to its original state or move to a new, more desirable state after being disturbed". SCRes refers to the ability of chains to prevent and withstand changes and recover the initial performance level after an unexpected disruption [22].

Much of the previous work on SCRes is related to quantifying the level of resilience before developing response and recovery [23–26]. In the wake of a disruptive event, significant participants in supply chains should be able to anticipate, be prepared and understand the extent of the impact caused by a disruption. They ought to generate strategies, so as to respond as quickly as possible to the impacts and redesign their resources to strengthen competencies and adapt to the resulting effects. Such modifications and the capacity for renewal to readapt allow for sound recovery from interruptions [25, 27]. The evaluation of SCRes in the literature has been dominated by theoretical and modeling approaches. The SCRes modeling and quantification approaches can be grouped under two different categories. The first category encompasses metric-based models that directly quantify (SCRes). In [23] five assessment metrics have been highlighted: economy, vulnerability, recovery time, uncertainty, and resilience-building considerations.

Other authors, such as [28–30] used performance loss to measure provider capacity and service performance. In reference to [31, 32] recovery time was used when assessing the impact of disruptions on a supply chain.

The second category of quantification models does not measure (SCRes) directly, but rather attempts to assess resilience strategies. Hirotsu et al. [33] assessed supply chain vulnerability using a multiple inputs and outputs model for different mitigation strategies. Ghomi-Avili et al. [34] measured supplier support and inventory levels as a resilience strategy to maintain performance during an outage. Hosseini and Ivanov, [35] evaluated (SCRes) by measuring supplier segregation, including geographically grouped suppliers.

However, despite the wide variety of theoretical studies which can be found, most emphasize the need for further empirical research on (SCRes), especially during outbreaks and disruptive events. Furthermore, previous literature on (SCRes) has mainly focused on exploring the facilitators, antecedents, common practices, capabilities and competencies needed to build resilient supply chains, without any emphasis on strategies to manage disruptions [36].

According to [25], these factors somehow form strategies to build (SCRes), and incorporate concepts such as the improvement of the supply chain in terms of promptness and better visibility, with the enhancement of flexibility, eliminating redundancies and optimizing the overall collaboration between partners. By doing so, it was possible to identify several SCRes strategies proposed in previous studies, which are summarized in **Tables 2** and **3**. Those are classified into two categories, the proactive and the reactive strategies.

Proactive strategies are primarily technology-oriented and rely on developing technical infrastructures such as digital connectivity and supply chain automation to prevent future disruptions [37, 38, 41]. Hosseini and Ivanov, [42] pointed out the location and regionalization of supply, while [51] delved into integrated approaches to supply chain risk management.

Proactive strategies	Description	References
Digital connection	The IOT and block chain Technology are proving to have great potential to boost SCRes through high connectivity, accuracy and transparency.	[37–40]
Supply chain automation	Regards the systematization of physical workflow and information throughout the supply chain for process improvement. This implies the use of technologies to minimize the dependence between people from different companies.	[37, 38, 41]
Supply chain regionalization/ location	Both supply and processing are located within the same region to meet local demand and reduce supply chain integrations. Consequently, risk is maintained within a specific region.	[42, 43]
Supply chain collaboration	Companies of all levels should work in synchronicity with each other in order to anticipate, predict and prevent potential threats to supply.	[36, 39, 43, 44]
Human capabilities	Involves human analysis of vast amounts of information and data, in which critical points are monitored and controlled in an entire supply chain system.	[36, 45]

Source: The author of this work adapted from [46].

Table 2.
Proactive supply chain risk mitigation strategies.

Reactive strategies	Description	References
Big data analytics (BDA) and real-time information systems	Supply chain information systems use BDA to collect, process and extract meaningful key insights from real-time data across a supply chain to support appropriate and timely decision-making strategies.	[46, 47]
Virtual markets	Involves the development of digital markets for the delivery of products and services.	[39, 48]
Test running	Test running has proved to be efficient in decision-making scenarios, as it deals with the uncertainty and complexity of problems related to supply chains.	[7, 38, 49]
Business continuity planning	Business continuity planning is extremely important when it comes to creating processes for the prevention of and recovery from potential disruptions in supply chain systems. (This can also be considered a proactive strategy.)	[27, 50]

Source: The author of this work adapted from [46].

Table 3.
Reactive supply chain risk mitigation strategies.

Other authors have proposed strategies related to a more social aspect, focusing on a social supply chain [42] whereas, [45] took to the subject of human capabilities.

Reactive strategies, on the other hand, mainly rely on a real-time information system and are based on decision-making with the use of data [46, 47], the creation of virtual markets in [48] and by means of supply chain simulation resources [38].

Some strategies, such as digital resource development and supply chain collaboration, can be both proactive and reactive, depending on the timing and purpose.

In a time as complicated as a pandemic, with huge variations in price, demand and consumer behavior, as shown in previous sections, developing a resilient supply chain is essential for a business to thrive.

4. Analysis and discussion

The previous sections of this article aimed at contextualizing the COVID-19 pandemic and its effects on supply chains. It also included data on consumer behavior, as well as certain strategies made to combat possible disruptions which took place in supply chains and how to counteract their effects in resilient way.

Among the supply chains shown, e-commerce stood out and happened to be an important escape valve at the beginning of the pandemic, which allowed some stores to remain in activity. More than 135,000 Brazilian stores adhered to e-commerce sales in the first three months of the pandemic in 2020, according to the ABComm.

Despite the advances of vaccinations and the reopening of businesses worldwide, online sales will continue to conquer space and become an excellent source of income for companies. For example, most customers are more likely to do their Christmas shopping online in 2021, as shown in **Table 4**. Yet, those figures indicate the sum of percentages combined and therefore exceeds 100% because products were purchased in different ways by the same customer. However, online shopping is always their first choice. This falls under the reactive strategy "developing digital marketplaces to deliver products and services" [39, 48].

Still on the topic of reactive strategies, in addition to those mentioned above, all strategies operate by means of technology, whether by test running, information systems or big data. It is crucial to elicit that the application of these strategies is of paramount importance, as they collect and process data in real time, and offer vital support to decision-making processes, preventing the mechanisms in a supply chain, from making wrong and hasty decisions that can affect the product flow, as well as the information exchange needed.

Undeniably, technology plays an important role in all stages of the supply chain. Consequently, the flow of information and goods is optimized, and disruptions may be prevented.

When it comes to food supply chains, all proactive strategies ought to be put to work under the best possible conditions. The use of technology is essential for a good flow within the chain, in terms of products or information. As there was an increase of 140% of food bought online in Europe in 2020, markets, which were able to successfully respond to that abrupt demand, gained significant competitive advantage over others.

On the other hand, as the movement of people was restricted, restaurants that depended exclusively on face-to-face services were severely affected and were forced to shut down. Under such conditions, bars and restaurants may target customers who enjoy eating out by investing in marketing strategies. As a result a large number of

The Internet	45%
Department stores	43%
Shopping malls	40%

Source: The author of this work.

Table 4.
The most popular purchasing methods for Christmas 2021.

people could be attracted to their establishments. However, the reactive strategies of these companies must be very well defined, as new variants of the coronavirus can emerge at any time, generating new waves of infections and local services may need to be closed. The regionalization of supply chains could have been adopted, among other proactive strategies, not only to mitigate the risk of disruptions during the pandemic, but also avoid future problems [42, 43]. When sourcing and processing are located within the same region, the risk remains within the same domain. As the pandemic started in China, a country where several supply chains branch out at their initial stage to other parts of the world, many other countries which had not yet suffered from the virus itself, were already heavily impacted due to the lack of supplies. If there was less dependence on this global hub, perhaps countries like Brazil, would be able to have a better response to the disruptions caused by the pandemic.

Still on the subject of technology, it is essential that all companies, not just bars and restaurants, adapt to the new reality of banking transactions. As most of the European population has become more knowledgeable about QR code technology, better purchase conditions have been offered to customers and therefore, Europe is more likely to prevent a breakdown in the last link of the supply chain.

Furthermore, the pandemic caused a change in consumer behavior due to disruptions in global supply chains. As a result, consumers have become more dependent on their closest communities. As fresh food, fruits and vegetables are delivered to them faster, people have adopted healthier eating habits. Moreover, by reducing consumption, avoiding waste and recycling materials, economy has become more circular.

5. Final considerations

This work aimed to present how the pandemic has affected and will continue to affect supply chains. It looked into the impacts caused by the pandemic, such as unexpected demand variations, changes in consumer behavior and how to mitigate sudden risks by using reactive strategies like e-commerce. That led to companies adapting their activities to a completely new reality of constant change and inconsistency.

The objective of this work was successfully achieved by researching data from reliable sources, acquiring a substantial amount of information to be analyzed, which not only allowed for a clear demonstration of the impacts on consumer behavior in Europe, North America and Brazil, but also to list strategies that were used to minimize risks, enabling companies to adapt to this new reality.

However, as it is a recent and ongoing issue, it was challenging to encounter more factual data regarding the effects in other continents, such as Africa and Oceania, in order to perform a truly global analysis.

As this work was completed in the first fortnight of December 2021, it was not possible to analyze and compare how Christmas shopping figures actually differed from the previous year. No compilation of data on European consumer behavior in 2021 was available, making it impossible to compare figures to those in 2020. This data can be explored in future work, with the compilation of information in future years, so as to confirm whether consumer behavior has really changed and set new trends, or if it was only temporarily induced by the incidence of the pandemic.

There is also a future perspective of several possibilities for achievable improvements to be implemented throughout the production chain.

Using the test-run method, it would be possible to develop a model to predict whether the trend of online shopping in supermarkets would continue. In other words, fewer consumers would physically go to shops and stores, which would make such locations redundant, and which could then be adapted to any sort of existing need. That would consequently reduce costs, allowing for investments to improve customer experience, from user interface to delivery and consumption.

Author details

Alexsandro Silva Solon* and Marcos Paulo Leite Silveira*
Universidade Federal do Triângulo Mineiro, Uberaba-MG, Brazil

*Address all correspondence to: alexsandro.solon@uftm.edu.br
and marcospaulo_silveira@hotmail.com

References

[1] Daim TU et al. Technology forecasting for wireless communication. Department of Engineering and Technology Management, Maseeh College of Engineering and Computer Science, Portland State University, Portland, OR 97201, USA, September 2008. DOI: doi:10.1016/j.technovation.2007.12.005

[2] Malhotra NK. Marketing Research: An Applied Orientation. 7th ed. Australia: Pearson; 2019

[3] Christopher M, Peck H. Building the Resilient Supply Chain. The International Journal of Logistics Management. 2005;**15**:1-13. DOI: 10.1108/09574090410700275

[4] Iyer K. Demand chain collaboration and operational performance: Role of IT analytic capability and environmental uncertainty. Journal of Business & Industrial Marketing. 2011;**26**(2):81-91. DOI: 10.1108/08858621111112267

[5] Lambert DM, Cooper MC. Issues in supply chain management. Industrial Marketing Management. 2000;**29**(1): 65-83. DOI: 10.1016/S0019-8501(99) 00113-3

[6] Handfield RB, Nichols EL Jr. Supply Chain Redesign: Converting Your Supply Chain into an Integrated Value System. USA: Prentice Hall; 2002. ISBN 0-13-060312-0

[7] Ivanov D, Das A. Coronavirus (COVID-19/SARS-CoV-2) and supply chain resilience: A research note. International Journal of Integrated Supply Management. 2020;**13**:90-102. DOI: 10.1504/IJISM.2020.107780

[8] National Academy of Medicine. 2020. Available from: https://nam.edu/publications/

[9] PwC. Annual Report 2020. Available from: https://www.pwc.com.br/pt/publicacoes/assets/2020/relatorio_anual_20.pdf

[10] Baldwin R, Tomiura E. Thinking ahead about the trade impact of COVID-19. In: Economics in the Time of COVID-19. London: Centre for Economic Policy Research; 2020. pp. 59-71

[11] OECD Economic Outlook. Volume 2020. Issue 2. DOI: 10.1787/39a88ab1-en

[12] Sharma R et al. Agriculture supply chain risks and COVID-19: Mitigation strategies and implications for the practitioners. International Journal of Logistics Research and Applications. 2020:1-27. DOI: 10.1080/13675567.2020.1830049

[13] The Financial Express, 2020. Available from: https://www.financialexpress.com

[14] Gazeta do Povo. Pandemia Covid-19 [Internet]. 2020. Available from: https://www.facebook.com/gazetadopovo/?hc_ref=ARTqUQie86vguR5qMm7LMrcuDBkTkX3bG6zHD4R_pr16sSOIEnJdY1H0pZBZEQ9WqRI&fref=nf&__xts__[0]=68.ARACNKy3eUsTs_QIPOTxbwFvK_JvaqwkfhuMxuXZJ3hyOgiRU6XlEMiwSkpTZPYi-RR5xgC_HmAFJCZhsHam6anj9O5Rdb6lkwFetDkO5WJAPUopmjvSvfdGFf9j8qwqbgCQRIDURIcdjG1R2TT8C4ZBqcRa9AMaQe_qBPqKuxbDlPVgkxvyH9n0ank3QQxj38-ALIQqrLOUn-gPb-hh-rhQ6JKKNkTYAbEQMCD54UvzPLJ9wivXb1RRLH76jb8JRf7zdE6qPB6wdvmjnljwGpVhoHhVsgUgXjtzgWQe0mnSc8oCcVbB

[15] BBC News, 2020. Available from: https://www.bbc.com/news/coronavirus

[16] Hobbs JE. Food supply chains during the COVID-19 pandemic. Canadian Journal of Agricultural Economics; 2020. DOI: 10.1111/cjag.12237

[17] Reply. Quentin, TD Replys, 2020. Análise do Impacto: Como a indústria de bens de consumo está respondendo ao coronavírus? Um relatório de dados sobre as mudanças no setor de FMCG relacionadas ao COVID-19. Brazil, 2020. Available from: https://www.reply.com/br/covid-19-fmcg-industry-impact-analysis

[18] European Central Bank. Annual Report, 2020. Available from: https://www.ecb.europa.eu/pub/pdf/annrep/ar2020~4960fb81ae.en.pdf

[19] Combe C. Introduction to e-Business: Management and Strategy. Elsevier; 2006, ISBN-13: 978-0-7506-6731-9

[20] CNDL – Confederação Nacional de Dirigentes Logistas. Panorama do Comércio 2021 – Brazil. Available from: https://site.cndl.org.br/panorama-do-comercio/

[21] Christopher M, Peck H. Building the Resilient Supply Chain. The International Journal of Logistics Management. 2005;**15**:1-13. DOI: 10.1108/09574090410700275

[22] Hendry DG et al. Value Sensitive Design: Shaping technology with Moral imagination. Cambridge, Massachusetts: MIT Press; 2019

[23] Hosseini S, Ivanov D, Dolgui A. Review of quantitative methods for supply chain resilience analysis. Transportation Research Part E: Logistics and Transportation Review. 2019;**125**:285-307. DOI: 10.1016/j.tre.2019.03.001

[24] Graveline N, Grémont M. Measuring and understanding the microeconomic resilience of businesses to lifeline service interruptions due to natural disasters. International Journal of Disaster Risk Reduction. 2017;**24**:526-538. DOI: 10.1016/j.ijdrr.2017.05.012

[25] Ivanov D, Dolgui A. Low-certainty-need (LCN) supply chains: A new perspective in managing disruption risks and resilience. International Journal of Production Research. 2018:5119-5136. DOI: 10.1080/00207543.2018.1521025

[26] Chowdhury MH, Quaddus M. Supply chain readiness, response and recovery for resilience. Supply Chain Management. 2016;**21**(6):709-731. DOI: 10.1108/SCM-12-2015-0463

[27] Hernantes J et al. Moving forward to disaster resilience: Perspectives on increasing resilience for future disasters. Technological Forecasting and Social Change. 2017;**121**:1-6. DOI: 10.1016/j.techfore.2017.05.01

[28] Ojha R et al. Bayesian network modelling for supply chain risk propagation. International Journal of Production Research. 2018;**56**:5795-5819. DOI: 10.1080/00207543.2018.1467059

[29] Torabi SA, Baghersad M, Mansouri SA. Resilient supplier selection and order allocation under operational and disruption risks. Transportation Research Part E: Logistics and Transportation Review. 2015;**79**:22-48, ISSN 1366-5545. DOI: 10.1016/j.tre.2015.03.005

[30] Schmitt AJ, Singh M. A quantitative analysis of disruption risk in a multi-echelon supply chain. International Journal of Production Economics. 2012;**139**(1):22-32, ISSN 0925-5273. DOI: 10.1016/j.ijpe.2012.01.004

[31] Ivanov D et al. Disruption-driven supply chain (re)planning and

performance impact assessment with consideration of pro-active and recovery policies. Transportation Research Part E: Logistics and Transportation Review. 2016;**90**:7-24. DOI: 10.1016/j.tre.2015.12.007. Disponível em: https://www.sciencedirect.com/science/article/abs/pii/S136655451500232X

[32] Simchi-Levi D et al. Identifying risks and mitigating disruptions in the automotive supply chain. Interfaces. 2015:375-390. DOI: 10.1287/inte.2015.0804

[33] Hirotsu Y, Maejima M, Shibusawa M, Nagakubo Y, Hosaka K, Amemiya K, et al. Comparison of automated SARS-CoV-2 antigen test for COVID-19 infection with quantitative RT-PCR using 313 nasopharyngeal swabs, including from seven serially followed patients. International Journal of Infectious Diseases. 2020;**99**:397-402, ISSN 1201-9712. DOI: 10.1016/j.ijid.2020.08.029

[34] Ghomi-Avili M, Naeini SGJ, Tavakkoli-Moghaddam R, Jabbarzadeh A. A fuzzy pricing model for a green competitive closed-loop supply chain network design in the presence of disruptions. Journal of Cleaner Production. 2018;**188**:425-442, ISSN 0959-6526. DOI: 10.1016/j.jclepro.2018.03.273

[35] Hosseini S, Ivanov D. Bayesian networks for supply chain risk, resilience and ripple effect analysis: A literature review. Expert Systems with Applications. 2020;**161**:113649, ISSN 0957-4174. DOI: 10.1016/j.eswa.2020.113649

[36] Tukamuhabwa B, Stevenson M, Busby J. Supply chain resilience in a developing country context: A case study on the interconnectedness of threats, strategies and outcomes. Supply Chain Management: An International Journal. 2017;**22**:486-505. DOI: 10.1108/SCM-02-2017-0059

[37] Ralston P, Blackhurst J. Industry 4.0 and resilience in the supply chain: A driver of capability enhancement or capability loss? International Journal of Production Research. 2020:5006-5019

[38] Hoffmann E et al. Supply chain management and Industry 4.0: Conducting research in the digital age. International Journal of Physical Distribution Logistics Management. 2019:945-955. DOI: 10.1108/IJPDLM-11-2019-399

[39] Kamble S, Gunasekaran A, Dhone NC. Industry 4.0 and lean manufacturing practices for sustainable organizational performance in Indian manufacturing companies. International Journal of Production Research. 2020;**58**(5):1319-1337

[40] Ivanov D, Dolgui A, Sokolov B. The impact of digital technology and Industry 4.0 on the ripple effect and supply chain risk analytics. International Journal of Production Research. 2019;**57**(3):829-846

[41] Tan WJ et al. Structural-aware simulation analysis of supply chain resilience. International Journal of Production Research. 2020;**58**(17):5175-5195. DOI: 10.1080/00207543.2019.1705421

[42] Iakovou E et al. Dual sourcing for mitigating humanitarian supply chain disruptions. Journal of Humanitarian Logistics and Supply Chain Management. 2014:245-264. DOI: 10.1108/JHLSCM-03-2013-0008

[43] Kochan CG, Nowicki DR. Supply chain resilience: A systematic literature review and typological framework.

International Journal of Physical Distribution & Logistics Management. 2019;**48**(8):842-865

[44] Barrane FZ, Ndubisi NO, Kamble S, Karuranga GE, Diane P. Building trust in multi-stakeholder collaborations for new product development in the digital transformation era. Benchmarking: An International Journal. 2020;**28**:205-228. DOI: 10.1108/BIJ-04-2020-0164

[45] Blackhurst J et al. An empirically derived agenda of critical research issues for managing supply-chain disruptions. International Journal of Production Research. 2007:4067-4081. DOI: 10.1080/00207540500151549

[46] Belhadi A, Kamble S, Jabbour CJC, Gunasekaran A, Ndubisi NO, Venkatesh M. Manufacturing and service supply chain resilience to the COVID-19 outbreak: Lessons learned from the automobile and airline industries. Technological Forecasting and Social Change. 2021;**163**:120447, ISSN 0040-1625. DOI: 10.1016/j.techfore.2020.120447

[47] Kamble S, Gunasekaran A. Industry 4.0 and lean manufacturing practices for sustainable organisational performance in Indian manufacturing companies. International Journal of Production Research. 2019:1-19

[48] Shariff H et al. A classification and selection model of e-marketplaces for better alignment of supply chains. Journal of Enterprise Information Management. 2006;**19**:483-503

[49] Kamble S, Gunasekaran A, Gawankar S. Sustainable industry 4.0 framework: A systematic literature review identifying the current trends and future perspectives. Process Safety and Environmental Protection. 2018;**117**. DOI: 10.1016/j.psep.2018.05.009

[50] Zsidisin G et al. An institutional theory perspective of business continuity planning for purchasing and supply management. International Journal of Production Research. 2005;**43**:3401-3420

[51] Chen L, Zhao X, Tang O, Price L, Zhang S, Zhu W. Supply chain collaboration for sustainability: A literature review and future research agenda. International Journal of Production Economics. 2017;**194**:73-87, ISSN 0925-5273. DOI: 10.1016/j.ijpe.2017.04.005

Chapter 8

Capacitated Clustering Models to Real-Life Applications

Marcos J. Negreiros, Nelson Maculan, Augusto W.C. Palhano, Albert E.F. Muritiba and Pablo L.F. Batista

Abstract

This chapter considers the use of different capacitated clustering problems and models that fits better in real-life applications such as household waste collection, IT teams layout in software factories, wholesales distribution, and staff's home collection or delivery to/from workplace. Each application is explored in its regular form as it is being developed by contractors and/or users. We consider for each application the aspects of solving the problem by the appropriate mathematical programming model and decision support methodology (using aggregated Geographical Information System and mobile technology) to hold correctly and most precisely the problems and difficulties related to instances in evaluation. The experience on these fields is here revealed in detailed form as the results obtained by using the techniques here explained.

Keywords: capacitated clustering problem, capacitated centerd clustering problem, mathematical modeling, decision support systems, optimization

1. Introduction

Clustering is a problem related to the process of assigning individuals to several disjoint partitions. Clustering problems can be classified in several ways. The methodology related to this study borns with the min-sum of squares clustering problem (MSSCP), [1, 2]. Its objective is to minimize the function of the least-squares distance between individuals to the geometric center of their partitions, such that the number of clusters to be built is known. This nonlinear binary problem is NP-Hard and can fit to several analytic applications such as classification, machine learning, partitioning, covering, location, and it is fully used by researchers to follow solutions to diverse type of important optimization problems linked with artificial intelligence (AI), [3–8].

To better understand the problem mathematically, suppose we have,

I: the set of individuals/items ($|I| = n$).

J: be the set of centers ($|J| = m$).

$|J| = p$: is the fixed number of *clusters*.

p: can also be the number of *clusters*.

L: is the set of the dimensions of the Euclidean space. $L = \{1, 2\}$ and $l = |L|$.

$a_i \in \mathbf{R}^l$: is the coordinate of an individual i.

$$x_{ij} = \begin{cases} 1, \text{if the individual } i \text{ is assigned to the center } j \\ 0, \text{otherwise} \end{cases}$$

$\bar{y}_j \in \mathbf{R}^l$: is the cluster center j.

The MSSCP can be formulated as:

$$(\textbf{MSSCP}) \ \textit{Minimize} \sum_{i \in I} \sum_{j \in J} \left\| a_i - \bar{y}_j \right\|_2^2 x_{ij} \ , \tag{1}$$

$$\text{Such that}: \sum_{j \in J} x_{ij} = 1, \ i \in I, \tag{2}$$

$$\bar{y}_j \in \mathbf{R}^l, \ j \in J, \ l \in L, \tag{3}$$

$$x_{ij} \in \{0, 1\}, \ i \in I, j \in J. \tag{4}$$

In the MSSCP model, the objective function (1) minimizes the total Euclidean distance from the items to their centers. The constraint (2) indicates the unique assignment of each item to a center of the cluster. Finally, constraints (3) and (4) refer to the decision variables. The MSSCP problem is NP-Hard, as indicated by Hansen & Jummar [3].

The continuous relaxation for *MSSCP* whose objective function is neither convex nor differentiable can be modified to be differentiable and convex.

The following mathematical development is in the work of [9], and it is a new formulation for the MSSC problem.

Let $a_{ik} \geq 0, \ k = 1, 2, \ldots, l$ be the coordinates of the given point $a_i \in \mathbf{R}^l, \ i \in I$. It is not difficult to consider a hyper-cube in $\mathbf{R}^l$, whose edge length is M, in which all points $a_i, \ \forall i \in I$ are included (M is the maximum dispersion between any two points of P).

We know that $x_{ij} \in \{ \ 0, 1 \ \}, \ i \in I, \ j \in J$, so we can write the following,

$$\left\| a_i - \bar{y}_j \right\|_2^2 x_{ij} = x_{ij}^2 \left(\sum_{k=1}^{l} \left(a_{ik} - \bar{y}_{jk} \right)^2 \right) = \sum_{k=1}^{l} x_{ij}^2 \left(a_{ik} - \bar{y}_{jk} \right)^2, \ i \in P, \ j \in J.$$

Consider $z_i \geq 0, i \in P$, we can write:

$$z_i + M(1 - x_{ij}) \geq \sum_{k=1}^{l} \left(a_{ik} - \bar{y}_{jk} \right)^2, \ i \in P, \ j \in J.$$

The MSSCP can be rewritten as:

$$(\text{MSSCP}') \ \text{Minimize} \sum_{i \in I} z_i \ , \tag{5}$$

$$\text{such that}: \sum_{k=1}^{l} \left(a_{ik} - \bar{y}_{jk} \right)^2 - M\left(1 - y_{ij}\right) - z_i \leq 0, \ i \in I, \ j \in J, \tag{6}$$

$$z_i \geq 0, \ i \in I, \ j \in J, \tag{7}$$

$$\sum_{j \in J} x_{ij} = 1, \ i \in I, \tag{8}$$

$$x_{ij} \in \{\ 0,1\ \},\ i \in I,\ j \in J, \tag{9}$$

$$\bar{y}_j \in \mathbf{R}^l_+,\ j \in J. \tag{10}$$

Quadratic constraints (6) form a convex set. This MSSC model can be implemented with success using XPRESS, CPLEX, and GUROBI optimization software because they already consider these constraints, then they can solve instances up to 1480 items, [10].

From location problems, the generalized p-median problem (gpMP) was introduced by Negreiros et al. [11] to be a kind of problem defining clusters by the individuals assigned to p-centers (items that are medians of the groups) that are closest to the sum-of-distances from the others of the same group. This new problem is linear and binary, and it is also NP-Hard. It is equally important in many analytic applications as the ones here developed.

For the Generalized p-Median Problem (gpMP) consider:

I: is the set of individuals.

J: is the set of possible medians (all individuals), $J \subseteq I$.

K: is the set of groups;

d_{ij}: is the shortest distance from individual i to its median j.

M_j : Auxiliary variable.

$$g_j^k = \begin{cases} 1, \text{if median } j \text{ used a cluster } k \text{ from the set of possible groups} \\ 0, \text{otherwise} \end{cases}$$

$$x_{ij} = \begin{cases} 1, \text{if an individual i is assigned to individual } j \text{ the median of the group} \\ 0, \text{otherwise} \end{cases}$$

The gpMP can be formulated as:

$$(\mathbf{gpMP})\ Minimize \sum_{i \in I} \sum_{j \in J} d_{ij}\ x_{ij}\ , \tag{11}$$

$$\text{Such that}: \sum_{j \in J} x_{ij} = 1,\ i \in I, \tag{12}$$

$$x_{ij} \leq x_{jj},\ i \in I, j \in J, \tag{13}$$

$$\sum_{i \in I} x_{ij} \geq M_j, j \in J, \tag{14}$$

$$\sum_{k \in K} g_j^k = M_j,\ j \in J, \tag{15}$$

$$g_j^k \in \{0,1\},\ j \in J,\ k \in K, \tag{16}$$

$$x_{ij} \in \{0,1\},\ i \in I,\ j \in J, \tag{17}$$

$$M_j \in \{0,1\},\ j \in J. \tag{18}$$

In the gpMP model, the objective function (11) minimizes the total distance from the items to their median centers that are items from the set of items. The constraint (12) indicates the unique assignment of each item to a median of the cluster.

Constraint (13) refers to the existence of a median if there is an assignment to it. Constraints (14) and (15) refer to a generalized assignment between items and their related median. In the constraints (16)–(18), we have the decision variables. The gpMP problem is also NP-Hard, [11].

A solution for both MSSCP and gpMP of a classical 50-item instance *German Towns* (**Figure 1a**) and four centers can be seen in **Figure 1b** and **c**. In both problems, we find the same structure of partitioning for clustering, but with different objectives and final solution representation. In the MSSCP, the final centers of the clusters are not part of the instance but with gpMP the centers are all part of the instance and the solution, **Figure 1(c)**.

The introduction of knapsack constraints to clustering problems was done by Mulvey & Beck [12] when they proposed the capacitated p-median problem CpMP. In this step, the clustering problems find a new research development segment in real applications. It is also a linear binary NP-Hard problem, because of this it became widely studied by many authors. In this problem, the capacity of each median can be distinct. The problem can be represented as follows:

I: is the set of individuals/items ($|I| = n$).

J: is the set of medians ($|J| = m$). $|J| = p$ is the fixed number of *clusters*.

p: is the given number of *clusters*/medians.

d_{ij} is the distance from individual i to its median j.

q_i: is the demand of an individual i.

Q_j: is the maximum capacity of median j.

$$x_{ij} = \begin{cases} 1, \text{if the individual } i \text{ is assigned to the median } j; \\ 0, \textit{otherwise}. \end{cases}$$

$$y_j = \begin{cases} 1, \text{if an individual } j \text{ is assigned to be a median} \\ 0, \text{otherwise.} \end{cases}$$

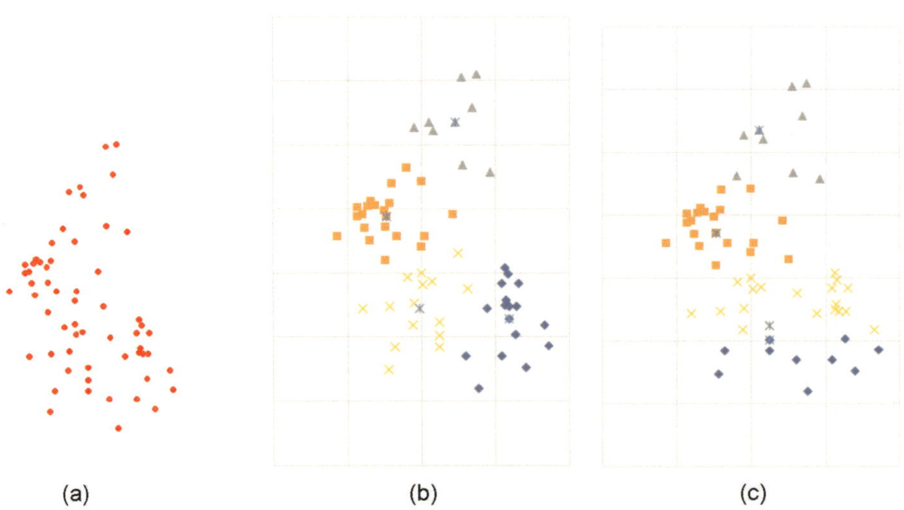

Figure 1.
Different clustering solutions of the same instance for distinct models and objectives. (a) Instance German Towns. (b) MSSC instance solution. (c) gpMP instance solution.

The CpMP can be formulated as:

$$(\mathbf{CpMP})\ \ Minimize \sum_{i\in I}\sum_{j\in J} d_{ij}\ x_{ij}\ , \tag{19}$$

$$\text{such that}: \sum_{j\in J} x_{ij} = 1,\ \ i\in I, \tag{20}$$

$$x_{ij} \le y_j,\quad i\in I, j\in J, \tag{21}$$

$$\sum_{i\in I} q_i\ x_{ij} \le Q_j\ y_j,\ \ j\in J, \tag{22}$$

$$\sum_{j\in J} y_j \le p\ , \tag{23}$$

$$y_j \in \{0,1\},\ \ j\in J, \tag{24}$$

$$x_{ij} \in \{0,1\},\quad i\in I,\ j\in J, \tag{25}$$

The objective function (19) minimizes the distance between *medians* and items assigned to each median. The constraint (20) defines that every item is to be assigned to exactly one median. The constraint (21) complements constraint (20) to indicate each item may be assigned to some other item that is used as a median. The constraint (22) considers the items demand may not overpass the limited capacity of the median they are assigned to. The constraint (23) limits the number of medians used from the set of medians. The constraints (24) and (25) refer to the decision variables of the problem.

At first it is considered that the set of medians are disjoint from the set of items, and as can be observed each median has its own maximum capacity, [12]. **Figure 2a** shows an instance of the CpMP where the set of items is in yellow squares ($q_i = 1, \forall i \in I$) and the set of medians appear in brown squares (maximum demand as shown in the label inside the square), $Q = [8, 7, 9, 11, 10]$ from top left to bottom right corner. In **Figure 2b**, we have the optimal solution of the model considering the situation proposed, [11].

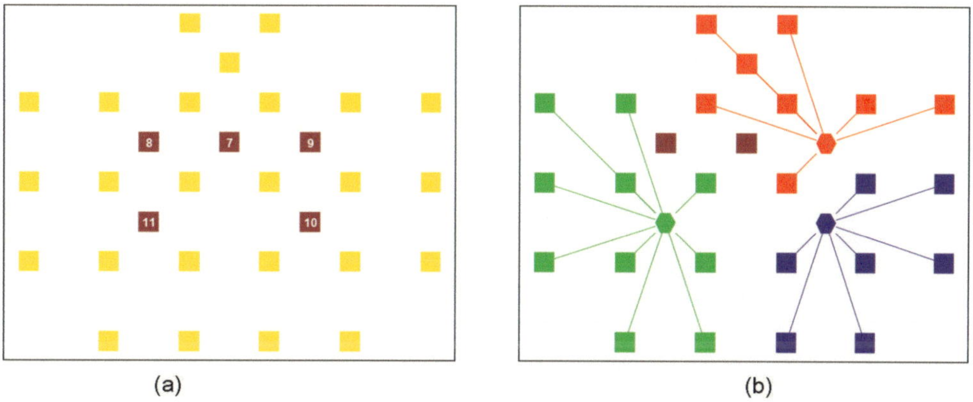

Figure 2.
CpMP instance and solution. (a) Capacitated p-median instance. (b) CpMP optimal solution of the instance (left).

If it is desired to extract medians from all the items by using different capacities, the different values of capacities may be repeated for each item in the set of medians. In this case, it is necessary to include a new set of constraints to avoid the use of more than one median between copies of the same item. The same model may be repeated, with the addition of this new constraint. **Figure 3a** shows the situation (instance) where there are no vertices previously defined as medians, instead all of them could be a median with capacity in $Q = [8, 7, 9, 11, 10]$. **Figure 3(b)** shows the result of the model obtained for this instance, [11].

If one wants to consider a direct model that solves the situation of selecting the best median from the set of items, once all the items can absorb all the capacities, a new model may be formulated, which we call generalized heterogeneous capacitated median problem (gHCMP). In this model, the variable $x_{jj} \in \{0, 1\}$ may represent the item that will be selected as the median from the set I, $(j \in I)$, it will use the appropriate assigned capacity. The order of the median is represented by the variable $g_j^k \in \{0, 1\}$, when it activates the capacity Q_k of median j assigned to group k. Now, it is not known *a priori* what is the assigned capacity for an item that will be chosen as median, the model decides by itself, [11].

For the Generalized Heterogeneous Capacitated Median Problem (gHCMP), consider:

I: is the set of individuals.

J: is the set of possible medians (all individuals).

K: is the set of groups.

d_{ij}: is the distance from individual i to its median j.

q_i: is the demand of individual i.

Q_k: is the maximum capacity of median k.

M_j, F_j: are auxiliary variables;

$$g_j^k = \begin{cases} 1, \text{if median } j \text{ used a cluster } k \text{ from the set of possible groups;} \\ 0, \text{otherwise} \end{cases}$$

$$x_{ij} = \begin{cases} 1, \text{if an individual } i \text{ is assigned to individual } j \text{ the median of the group;} \\ 0, \text{otherwise} \end{cases}$$

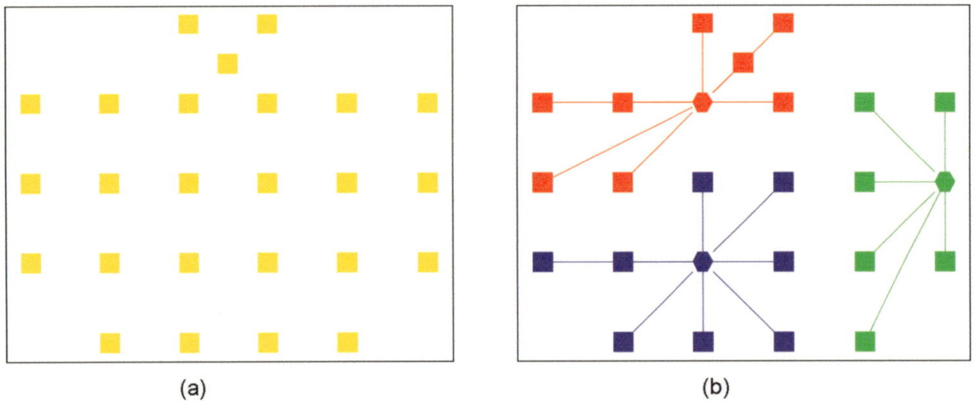

Figure 3.
gHCMP instance and solution. (a) A generalized heterogeneous capacitated median instance. (b) gHCMP optimal solution.

The gHCMP can be formulated as:

$$(\textbf{gHCMP})\ \textit{Minimize} \sum_{i \in I} \sum_{j \in I} d_{ij}\ x_{ij}\ , \tag{26}$$

$$\text{such that}: \sum_{j \in I} x_{ij} = 1,\ i \in I, \tag{27}$$

$$x_{ij} \le x_{jj},\ i \in I,\ j \in I, \tag{28}$$

$$\sum_{i \in I} x_{ij} \ge M_j,\ j \in J, \tag{29}$$

$$\sum_{k \subset K} g_j^k = M_j,\ j \in J, \tag{30}$$

$$\sum_{j \in J} g_j^k = 1,\ k \in K, \tag{31}$$

$$\sum_{i \in I} q_i\ x_{ij} = F_j,\ j \in J, \tag{32}$$

$$F_j \le \sum_{k \in K} Q_k\ g_j^k,\ j \in J, \tag{33}$$

$$g_j^k \in \{0,1\},\ j \in J,\ k \in K, \tag{34}$$

$$x_{ij} \in \{0,1\},\ i \in I,\ j \in J, \tag{35}$$

$$M_j \in \{0,1\},\ F_j \ge 0,\ j \in J. \tag{36}$$

The objective function (26) minimizes the distance between *medians* and each item assigned to its median. The constraint (27) defines that every item has to be assigned to exactly one median. The constraint (28) defines that once a median is used an item can be assigned to it. The constraint (29) joins items to medians sets and medians to groups sets. The constraints (30) and (31) warrant that each selected median assigns just one group. The constraint (32) considers the sum of items demand assigned to a median may generate a forward capacity. The constraint (33) indicates the forward capacity for each selected median may not overpass the limited capacity of its assigned group. The constraints (34)–(36) refer to the decision variables of the problem.

The gHCMP problem is NP-Hard as CpMP, [12].

The capacitated centered clustering problem (CCCP) was proposed by [13], considering a constrained process of min-sum clustering for a set of demanding individuals on the Euclidean space. Once clustering can be performed in various ways, this problem searches for solutions where limits are present on the capacity of the clusters or even their size (maximum number of individuals per cluster). This problem is also NP-hard and introduces interesting research topics for exploring combinatorial optimization methods for solving it, [14].

Two problems were introduced by [13]: *p*CCCP and generalized CCCP (*g*CCCP). In the first type, the capacitated constrained min-sum clustering is bounded by several groups. Meanwhile, in the second generalized type, there is no limit on the number of groups; however, a fixed cost is added to the objective function to open a new cluster. To evaluate these problems, [13] prepared a set of test instances extracted from different sources (wholesale distribution, garbage collection, Euclidean TSP, and capacitated p-median problem).

The literature explores several applications, such as in dry food distribution logistics, designing zones for urban garbage collection, the territorial design of sales, tropical disease control (e.g., dengue, chikungunya, zika, and yellow fever), agro-food supply chain network, vehicle routing, bulk milk collection, routing subscribers of newspapers, integrating commercial and residential pickup and delivery networks, allocating emergency shelters to protect the local population during possible natural disasters, computational biology, facility location, offshore wind farm electrical cable layout, and IT teams placement on the floor of software factories, [11, 13–22].

In many of these applications, the pCCCP was also adopted as a sub-problem to solve location-routing problems, capacitated clustering problems, districting, substation placement problems, and capacitated and periodic vehicle routing [20–26].

In the following sections, we consider four applications: factory floors IT-Teams layout in Section 2, logistic (wholesales, staff collection and dispatching) in Section 3, health (EMS and dengue prevention, control, and combat) in Section 4 and waste collection (domiciliar, selective, street sweeping, hill climbing) in Section 5. All these contexts CCP are applied with different forms for each application. It is discussed in detail the models proposed, it is considered the formulation and complexity of the appropriate CCP model used. For some of the approaches we discuss decision support systems (DSS) we developed and we also include forward steps as vehicle routing methodologies to complete the operational level. For the end, we include our experience with these models and DSS technology developed by us revealing the optimization results we achieved in applications such as: IT-Teams, staff collection/dispatching, and waste collection (domiciliary and human street-sweeping).

2. Layout in factory floors

Following we consider an important organizational problem that is performed weekly in software factories, during the development of software projects by teams in a software factory. The relevance of the CCP in this context was confirmed in two major Brazilian software factories: DATAPREV and SERPRO by [11].

2.1 IT-teams layout in software factories

Just before and during the pandemic of COVID-19, the software factories were managing their production by home office. This intent changed the behavior of development teams and the interactions between their members. The living costs highly dropped for factories but increased to their members (energy, food, etc.) although the quality of life improved for most of them. But it confirmed that some management problems may arise when dealing with dispersed teams: ensuring control in completing critical tasks; supporting a feature that needs to be updated regularly; always have some member of the team when a failure is detected or a new task need to be implemented and others. Actual availability of very high-speed internet and impressive software tools for communication by voice, video (alive and recording) and data transfer interchange, between groups, the concept of agile development used in software factories in Brazil changed most recently as labor regulations to deal with home office.

Home office and in-person office labor mixed structure is now evident in major software factories, maintaining the obligation of regular meetings of teams in the factory floor.

Schedules that alter the contingent of projects, or the opening and closing of projects, can motivate companies to move members between workstations, or even teams between rooms or floors of the factory. When this movement is done without criteria, it consequently disorganizes teams, separating members in dispersed places configuring undesirable circumstances such as those already mentioned. The definition of the layout of the teams can be complex in companies that have a high project turnover rate and a large number of employees.

It is explained here a real problem Information Technology-Teams (IT-Teams) project placement, from Brazilian federal company Dataprev. This IT company operates as a software factory and relies on standardized development teams skilled in different contexts.

Each project team has a number of people previously defined by managers in accordance with the complexity or work effort estimated to the project. This number is usually different between teams due to the different complexities of projects.

Dataprev's project teams are located in a large hall of a building, where workstations are arranged in cells composed up to four stations. There were 121 workstations where teams are allotted, see **Figure 4**.

In this context, it is a common problem to structure the layout of the factory workstations in such a way that the distinct groups of the projects may maintain their own members close to each other. There are many ways of building groups with individuals of any kind and characteristic. Although humans are specialized in separating things to perform better decisions when partitions are to be the expected result, this decision process is not possible when the number of individuals is high. When the individual have a demand/offer and a limited capacity is involved while building the groups, the resulting decision problem is more difficult because of its combinatorial counterpart. Also, in this context, there are many ways of doing *clusters* of IT-Teams.

If one wants to consider a direct model that solves the situation of selecting the best median from the set of items, once all the items can absorb all the capacities, the appropriate model to be used is the generalized heterogeneous capacitated median problem (gHCMP), previously defined.

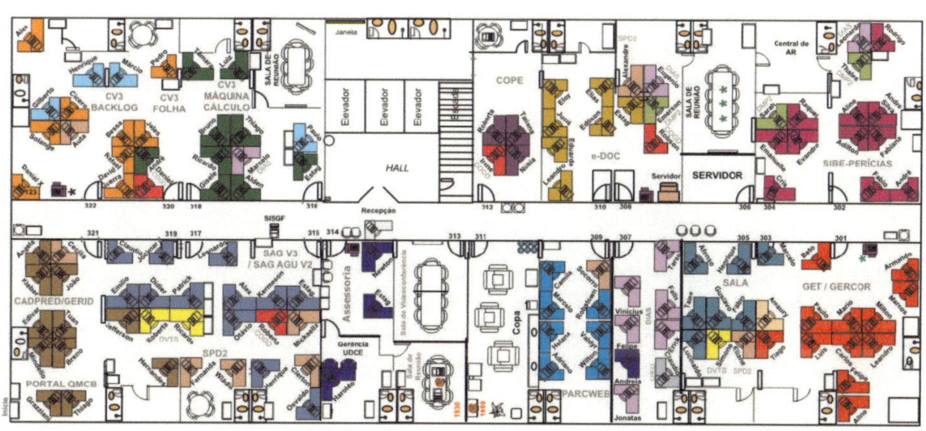

Figure 4.
Dataprev enterprise IT-Teams layout.

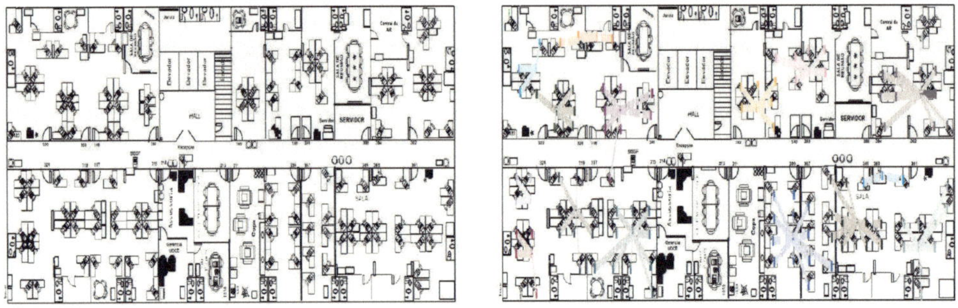

(a) (b)

Figure 5.
Process to layout IT-Teams in Dataprev factory floor. (a) An instance to layout Dataprev IT-Teams. (b) Optimal solution to a Dataprev layout instance.

In **Figure 5(a)**, it is presented an instance to layout workstations (points) in Dataprev factory floor at Fortaleza/CE, and in **Figure 5(b)**, it is presented the solution of the model gHCMP, which can be used to solve the factory layout problem.

In fact, the process of layout is more dynamic than what is proposed. Projects end and new projects arise in software factories. The management of this available staff is another important problem that arises in the factory. For the layout, in fact it does not move, because the new instance will run only over the workstations available. In fact, this affect the general layout, and reconfigurations may be needed with minimum change between actual layout.

3. Logistic

There are many problems in logistic that consider the use of decision support systems to support strategic, tactical, and operational level of planning. These systems appear most recently with great success in distribution. Most of them were designed by using classical routing algorithms, but the most succeeded went to the area of clustering, once they carry out more information than only distance to join customers. Aspects related to the customers' attributes (size, type of business, importance, etc) are relevant for many distribution companies to prepare service plans oriented.

Following we discuss two problems related to human force: first in labor (sales force distribution) and the second for the worker (staff collection and dispatching).

3.1 Wholesales distribution

In wholesales distribution, actually practiced in Brazilian major cities, the retailer owns its sales-force that is disperse in major cities boroughs, commercial corridors, or downtown. There are several different types of goods to sell to many stores or groceries. Actually the salesmen are equipped with mobile systems that are used to obtain the requests and/or propose offers to special customers along their area.

The market can be efficiently covered by:

1. Aggregating customers by their similarity in product consumption, revenues, and distance proximity;

2. Aggregating by the number of customers per cluster.

The first aggregation process abovementioned is related to the attributes of each customer of the market. Then it can be defined a specialized salesman to cover a market accordingly to the product consumption.

The second is the constraint associated with the capacity to cover the market by a salesman. This division is to define a workload of a day for a salesman or for the sales-force.

It is common in certain markets that the sales-force can cover the market customers' in a week. Or periodically for some customers. The periodic time associated with a customer depends on its relevance to the revenues or the type of product associated with its market share.

The sales-force can be limited to the number of salesman, and it may be required to be calculated the necessary salesmen to cover the areas considering the proposed workload capacity and attributes requirements.

Bard & Jarrah [19] conducted a research on network design focused in a division that considers commercial and residential customers separately, situation motivated by their respective geographic densities and the size and frequency of their demand. To construct driver work areas, the authors indicated that it is necessary to consider the expected demand, vehicle capacity, time on the road, and aspect ratio of individual territories. This triggers a capacitated clustering problem with side constraints. Based on a previously developed column generation algorithm, a case study was conducted involving several scenarios that integrated two networks, to determine the extent of the resulting benefits. The analysis indicated that a significant reduction in fleet size could be achieved when the two networks were either partially or fully combined. It also demonstrated that small reductions with respect to practice were possible when they were separately maintained, with the added benefit that the geometry of the resultant clusters satisfies certain desirable properties in terms of their contours and aspect ratios.

The appropriate model to solve this type of problem was proposed by Negreiros & Palhano [13] with CCCP. The general version is more appropriate when the sales-force needs to be known, and the single version is appropriate to the case when the sales-force dimension is known in advance.

3.1.1 A novel periodic capacitated clustering problem model

A new and more general model can be proposed to hold the periodic characteristics above mentioned. Let us suppose that the generalized periodic CCCP can be represented using the following set of parameters and variables:

l: dimension of the Euclidean space ($l = 2$, in our case).

L: set of dimensions of the Euclidean space ($|L| = l$, in our case).

I: set of individuals.

J: set of clusters.

H: set of time horizon, $H = \{h_1, h_2, \dots, h_T\}$.

a_i: vector of l dimension with the coordinates of the individual i.

T: period of time horizon.

f_i: frequency of an individual i in T.

tbv_i: minimum time between visits of an individual i during time horizon.

Q_j^h: maximum capacity of a cluster j in time horizon h.

q_i^h: demand of an individual i in time horizon h.

n_j^h: number of individuals in the cluster j in time horizon h.

$\overline{x}_j^h$: vector of dimension l representing the coordinates of the *cluster* center j in time horizon h.

$$y_{ij}^h = \begin{cases} 1, \text{if individual } i \text{ is assigned to cluster } j \text{ in time horizon } h \\ 0, \text{otherwise} \end{cases}$$

We consider the generalized periodic capacitated centerd clustering problem g PCCCP as follows:

$$(\text{gPCCCP})\text{Minimize} \sum_{i \in I} \sum_{j \in J} \sum_{h \in H} \| a_i - \overline{x}_j^h \| \; y_{ij}^h \; , \tag{37}$$

$$\text{such that} : \sum_{j \in J} y_{ij}^h = 1, i \in I, h \in H, \tag{38}$$

$$\sum_{i \in I} y_{ij}^h = n_j^h, j \in J, h \in H, \tag{39}$$

$$\sum_{i \in I} a_i \; y_{ij}^h = n_j^h \overline{x}_j^h, j \in J, h \in H, \tag{40}$$

$$\sum_{i \in I} q_i^h \; y_{ij}^h \leq Q_j^h, j \in J, h \in H, \tag{41}$$

$$\sum_{j \in J} \sum_{h \in H} y_{ij}^h = f_i, i \in I, \tag{42}$$

$$\sum_{j \in J} \sum_{h \in H} h \; y_{ij}^h \geq (f_i - 1) \; tbv_i, i \in I, \tag{43}$$

$$\overline{x}_j^h \in \mathbf{R}^l, n_j^h \in \mathbf{N}, y_{ij}^h \in \{0, 1\}, \quad \forall i \in I, \forall j \in J, \forall h \in H. \tag{44}$$

The objective function (37) minimizes the Euclidean distance among individuals assigned to the same cluster in horizon h. It is also possible to use min-sum-of-square measure (minimize the total variance of the clusters). Constraint (38) assigns each individual to exactly one cluster in time horizon h. Constraint (39) assigns the number of individuals per *cluster* (n_j^h) in time horizon h. Constraint (40) defines the centers of each *cluster* (x_j^n) in time horizon h. For each cluster center j in time horizon h, constraint (41) limits the sum of the demands of its assigned individuals. Constraint (42) refers to state the necessary frequency of the individuals during time horizon T. Constraint (43) guarantees the interval between visits on the intervals for the same customer. Constraint (44) refers to the domain of the decision variables.

The gPCCCP is nonlinear with linear constraints (38), (39) and, (41) and nonlinear constraint (40). Its difficulty (NP-Hardness) can be found in the the center location constraints (39) and (40), knapsack constraint in (41), the periodic component in (42) and (43), and the partitioning process encapsulated in the problem.

The continuous relaxation for gPCCCP, whose objective function is neither convex nor differentiable, can be modified to be differentiable and convex.

For the nonlinear constraint (40), we can linearize them using more 0–1 variables, as follows.

$$\sum_{i\in I} y_{ij}^h = n_j^h,\ j\in J,\ h\in H,$$

and

$$\sum_{i\in I} a_i y_{ij}^h = n_j^h \overline{x}_j^h,\ j\in J,\ h\in H.$$

implying

$$\sum_{i\in I} a_i y_{ij}^h = \left(\sum_{i\in I} y_{ij}^h\right)\overline{x}_j^h,\ j\in J,\ h\in H.$$

or

$$\sum_{i\in I} a_i y_{ij}^h = \sum_{i\in I} y_{ij}^h \overline{x}_j^h,\ j\in J,\ h\in H.$$

$$\overline{x}_j^h = \left(\ \overline{x}_{j,1}^h\ \overline{x}_{j,2}^h \ldots \overline{x}_{j,l}^h\ \right)^T.$$

$$\text{We have to compute } y_{ij}^h \overline{x}_{j,r}^h,\ r\in L.$$

$$\text{We replace :}$$

$$y_{ij}^h \overline{x}_{jr}^h \quad \text{by} \quad z_{ijr}^h,\quad r\in L.$$

It is very easy to obtain lower bound $LB = -M$ and upper bound $UB = M > 0$ for each $\overline{x}_{jr}^h$.

$$-My_{ij}^h \le z_{ijr}^h \le My_{ij}^h,\ r\in L$$
$$-M\left(1-y_{ij}^h\right) + \overline{x}_{jr}^h \le z_{ijr}^h \le \overline{x}_{jr}^h + M\left(1-y_{ij}^h\right),\ r\in L.$$

Therefore, we can rewrite problem gPCCCP as follows.

$$(gPCCCP')\quad \text{Minimize} \sum_{i\in I}\sum_{j\in J}\sum_{h\in H} w_{ij}^h\ , \tag{45}$$

$$\text{Such that,}$$

$$\left(w_{ij}^h\right)^2 \ge \sum_{r=1}^{l}\left(m_{ijr}^h\right)^2,\quad w_{ij}^h \ge 0,\qquad i\in I,\ j\in J,\ h\in H, \tag{46}$$

$$-M\left(1-y_{ij}^h\right) + \left(a_{ir} - \overline{x}_{jr}^h\right) \le m_{ijr}^h, r\in L,\ i\in I,\ j\in J,\ h\in H, \tag{47}$$

$$m_{ijr}^h \le \left(a_{ir} - \overline{x}_{jr}^h\right) + M\left(1-y_{ij}^h\right), r\in L,\ i\in I,\ j\in J,\ h\in H, \tag{48}$$

$$\sum_{j\in J} y_{ij}^h = 1,\ i\in I,\ h\in H, \tag{49}$$

$$\sum_{i\in I} a_{ir} y_{ij}^h = \sum_{i\in I} z_{ijr}^h, r\in L,\ j\in J,\ h\in H, \tag{50}$$

$$-My_{ij}^{h} \le z_{ijr}^{h} \le My_{ij}^{h}, \quad r \in L, \ i \in I, \ j \in J, \ h \in H, \tag{51}$$

$$-M\left(1 - y_{ij}^{h}\right) + \overline{x}_{jr}^{h} \le z_{ijr}^{h}, \quad r \in L, \ i \in I, \ j \in J, \ h \in H. \tag{52}$$

$$z_{ijr}^{h} \le \overline{x}_{jr}^{h} + M\left(1 - y_{ij}^{h}\right), \quad r \in L, \ i \in I, \ j \in J, \ h \in H, \tag{53}$$

$$\sum_{i \in I} q_i^h y_{ij}^h \le Q_j^h, \quad j \in J, \ h \in H, \tag{54}$$

$$\sum_{j \in J} \sum_{h \in H} y_{ij}^h = f_i, \quad i \in I, \tag{55}$$

$$\sum_{j \in J} \sum_{h \in H} h \ y_{ij}^h = (f_i - 1) tbv_i, \quad i \in I, \tag{56}$$

$$\overline{x}_{jr}^{h} = \left(\ \overline{x}_{j1}^{h} \ \overline{x}_{j2}^{h} \ldots \overline{x}_{jl}^{h} \right)^{T} \in \mathbf{R}^{l}, \ j \in J, \ h \in H, \tag{57}$$

$$y_{ij}^h \in \{0,1\}, \ z_{ijr}^h \in \mathbf{R}, {}^{h}_{ijr} \in \mathbf{R}, \quad i \in I, \ j \in J, \ h \in H, \tag{58}$$

$$a_{ir} = (\ a_{i1} \ a_{i2} \ldots \ a_{il} \ )^{T} \ . \tag{59}$$

Constraint (46) is second-order cone, which forms a convex set.

3.1.2 A spatial DSS to gPCCCP

Graphvs™ developed a new software tool that can hold a preliminary version of this problem. In the DSS, SisRot FULL®, the problem solved considers the partition of the customers for salesmen in two steps: first division of the coverage considering the workload for the total time horizon, and second division of the coverage assigned to each salesman along the time horizon. In this approach, the frequency to visit a customer is unitary. The steps of the solution can be followed in **Figures 6–8**, where in **Figure 6a**, it is seen all the customers to be visited in an area, in **Figure 6a**, it is seen the customers assignment to salesmen. In **Figure 6b**, it is seen the division of the area to be covered along 5 days of a week for two salesmen. In **Figure 8a**, it is seen the

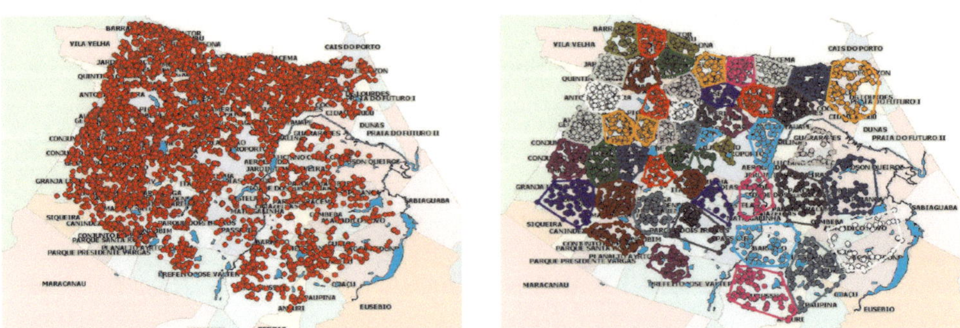

(a) (b)

Figure 6.
Process t layout IT-teams in Dataprev factory floor. Step 1 to build a clustering solution to sales force. (a) Customers distributor position spreading in the city of Fortaleza/CE. (b) Covering areas of 50 salesmen.

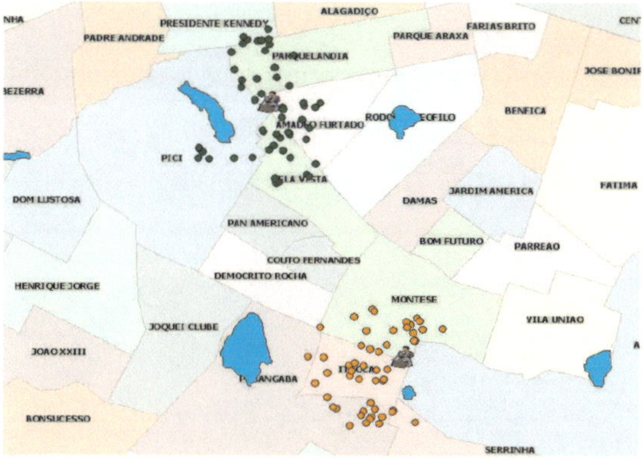

Figure 7.
The customers per week of two salesmen.

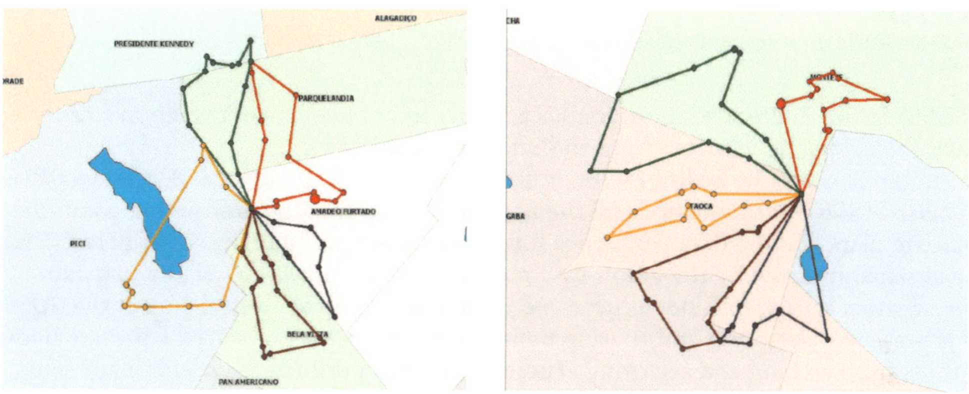

(a) (b)

Figure 8.
Step 2 building the routes per week of two salesmen. (a) Week covering routes of salesman 1. (b) Week covering routes of salesman 2.

thematic routes (sequence of visiting customers) for each day of the week for salesman 1 and in **Figure 8b** for salesman 2.

3.1.3 Structuring solutions in a spatial DSS

A tool to manage all the information and decisions into a spatial decision support system is described by its architecture in the **Figure 9**. To better trace the solution and assign to the decision process the solution to the sales-force in the next step. With the solution, the assigned dispatch went to the selling system that deploys the routes of customers' visits to the salesmen.

In the SDSS exposed in **Figure 9**, it is presented the contents of a framework of software that work in conjunction to deal with the problem of planning and organizing the salesmen force of a distribution company. The system is implemented in a cloud unity (VPS) where its components can share information of its competence. The major system is composed of two parts, the first is the spatial decision support system

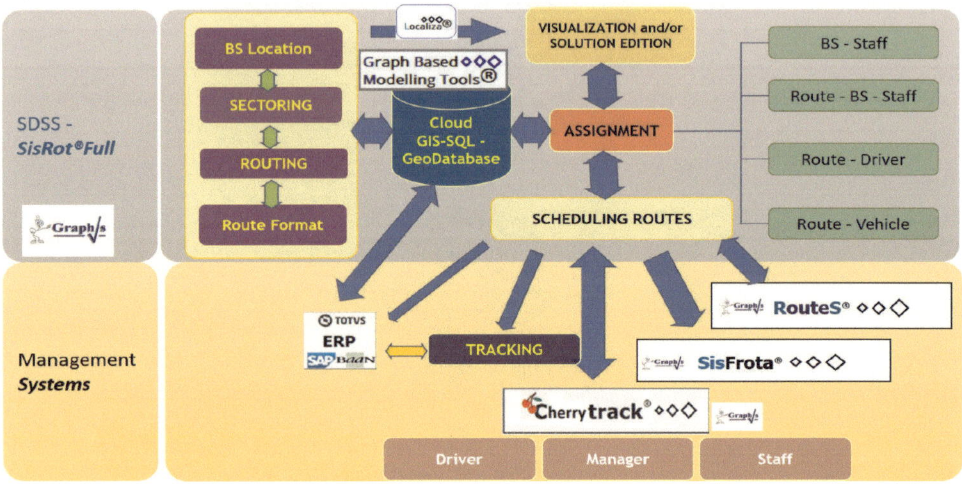

Figure 9.
SDSS for Wholesales planning distribution of salesmen force.

(SDSS) SisRot Full®, and the second is a set of tools to manage information from sales, sales force, and vehicles' tracking and salesmen monitoring.

In the SDSS of the SisRot Full®, it has embedded a GIS and an SQL-Geodatabase enclosed (ESRI™) that include all the necessary geographical information about the place to plan, and the data from sales force and its constraints. The street network can be also manipulated by the graph base modeling tools (GBMS) available that can revise street status, forbidden turns and directions. As an analytical part of the SDSS, it is presented sectoring and routing tools that contains a set of state-of-the-art algorithms from routing and sectoring. These algorithms perform the preliminary solutions that can be edited using productions available to move customers between routes, reassign routes to new depots, create or delete routes, and many other. The final solution can be sent to be viewed in different map tools such as Google Maps™ and Here Maps™ by using the software Routes®, to better evaluate the routes with traffic conditions during their run. The final routes solution can be also assigned to the salesmen (or driver) and can be scheduled accordingly the days of covering the salesman's set of customers.

External software for managing the maintenance of vehicles can be included such as SisFrota®, and managing sales force such as ERPs from Baan, Totvs, SAP, and many other. Graphvs™ most recently introduced a new tool to follow salesman while in the field, it is called CherryTrack®. In this tool it is possible to follow any salesman simultaneously and its results while they are selling using the manager tool, and it can give to the salesman his route while he/she is walking/driving through the customers assigned to the day of the work, including all the information about his results per customer. The mobile CherryTrack Manager and Driver tools are developed to Android and make great difference in performing and controlling the sales force in the field.

The solution abovementioned does not include a more general frequency and heterogeneous capacity (workload) between salesmen. Although we propose a more general model, our proposition open new research opportunities for this application that can be also fitted to Hospitaller Waste Collection, Inventory Routing, and many other.

3.2 Staff collection and dispatching

In the staff collection and dispatching problem, a company is contracted to pickup and deliver the staff of the firm accordingly their turns using a fleet of special busses, and considering that each trip does not exceed a maximum of time spend on traveling (2 h) and the hour to arrive and leave from the workplace. There is a series of different problems to solve, but location-routing and scheduling are the most common used path to solve the problem.

In fact, the preliminary problem to solve is to locate bus-stops close to the staff's homes (at least 800 m far, by Brazilian law) and second to perform routes that pass through the stops considering the capacity of the busses available, maximum time, and other possible existing constraints. In fact the location-routing can be developed in just one run. This problem was studied preliminarly by Madsen [27], Laporte et al. [28], Laporte [29, 30], and Borges et al. [31]. It is being studied from the literature mostly for school bus routing and scheduling, when there is a necessity to pickup and delivery students in located bus-stops, [32, 33].

The use of original distance constrained and capacitated p Median Problem DCpMP for the staff collecting and dispatching problem is a new form to understand this problem, it can be defined in two stages for aggregating staffs to bus-stops close to their homes and aggregate bus-stops to vehicles accordingly the staffs' turns. The CpMP problem was first proposed by Mulvey & Beck [12], here it is modified to be DCpMP to define the placement of the bus-stops as follows:

I: be the set of individuals/items ($|I| = n$).
J: be the set of bus stops - medians ($|J| = m$).
$|J|$: is the fixed number of *clusters*.
p: can also be the number of *clusters*/medians.
d_{ij}: is the distance from individual i to its median j.
D_{max}: is the maximum distance from staff's home and its median.
q_i: is the demand of an individual i.
Q_j: is the maximum capacity of median j.

$$x_{ij} = \begin{cases} 1, \text{if the individual } i \text{ is assigned to the median } j \\ 0, \text{otherwise} \end{cases}$$

$$y_j = \begin{cases} 1, \text{if an individual } j \text{ is assigned to be a median} \\ 0, \text{otherwise} \end{cases}$$

The DCpMP can be formulated as:

$$(\textbf{DCpMP})\ Minimize \sum_{i \in I} \sum_{j \in J} d_{ij} x_{ij}\ , \tag{60}$$

$$\text{such that}: \sum_{j \in J} x_{ij} = 1, \quad i \in I, \tag{61}$$

$$x_{ij} \leq y_j, \quad i \in I, j \in J, \tag{62}$$

$$\sum_{i \in I} q_i\ x_{ij} \leq Q_j\ y_j, \quad j \in J, \tag{63}$$

$$d_{ij}\ x_{ij} \leq D_{max}\ y_j, \quad i \in I,\ j \in J, \tag{64}$$

$$\sum_{j \in J} y_j \leq p, \tag{65}$$

$$y_j \in \{0, 1\}, \quad j \in J, \tag{66}$$

$$x_{ij} \in \{0, 1\}, \quad i \in I, j \in J. \tag{67}$$

The objective function (60) minimizes the distance between *medians*/bus-stops and staff assigned to each median. The constraint (61) defines that every staff is to be assigned to exactly one median. The constraint (62) complements constraint (61) to indicate each staff may be assigned to some other bus-stop. The constraint (63) considers the number of staff may not overpass the limited capacity of the bus-stop they are assigned to. The constraint (64) considers the staff i can only be assigned to bus-stop j if its distance is less than the maximum distance. The constraint (65) limits the number of bus-stops used from the set of bus-stops. The constraints (66) and (67) refer to the decision variables of the problem.

Figure 10a and **b** present the first step of the problem in the city of Fortaleza/CE, in two situations: The geographical spread of the 79 employees of a local distributor for a given turn and the candidate bus-stops. The solution is the selected bus-stops as can be seen in **Figure 11a.** In more detail, we have the assignment of the employees to

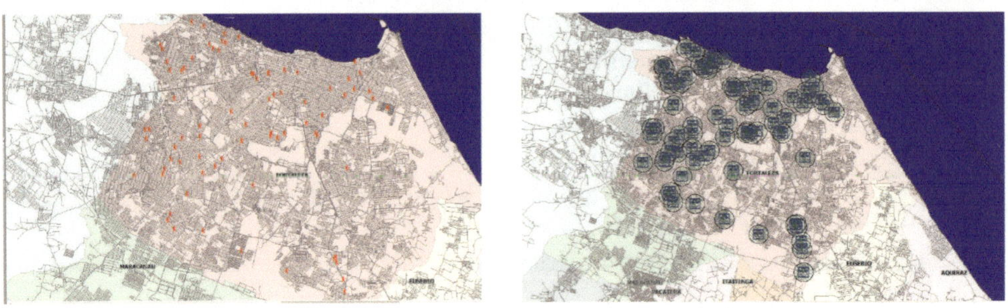

(a) (b)

Figure 10.
Homes distribution of the staff and candidate bus-stops. (a) Staff position for Staff Collection & Dispatching Problem. (b) Candidate bus-stops for Staff Collection & Dispatching Problem.

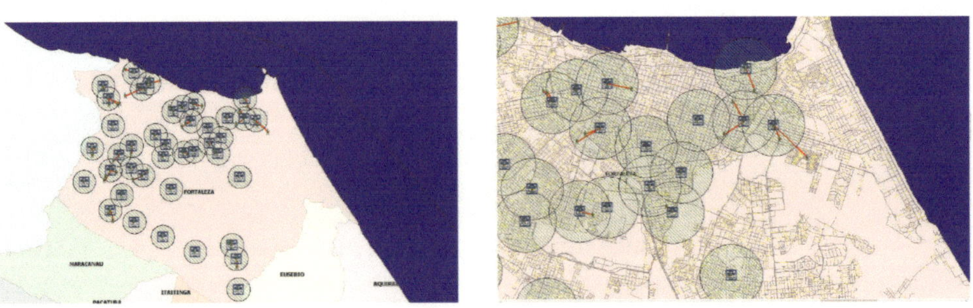

(a) (b)

Figure 11.
First phase of the staff collection and dispatching problem. (a) Bus-stops selection and assignment to each staff's home. (b) Detailed assignment of staff to bus stops.

Figure 12.
Second and third phase of the staff collection and dispatching problem. (a) Aggregation of bus-stops for the Staff Collection & Dispatching Problem. (b) Final routes to the Staff Collection & Dispatching Problem.

their closest stops in **Figure 11b**. It is also noted that the final solution can violate or not the constraints of the problem, here it is seen that some employees are out from the 800 m distance. In this case it is to the expert the preference to move to the next bus-stop, or to indicate the worker has to take another way to go/back to work.

The second problem is to produce aggregated stops to perform the final bus routes. As can be seen, **Figure 12a** presents the routes of the busses between stops and in more detail the employees moving to their stops in **Figure 12b**.

Defining HCCCP as the basic model in this step, it is possible to build high-quality routes. A post-optimization procedure can be performed to achieve time constraints, which are not considered in the clustering steps. Applying an appropriate metaheuristic such as GRASP+PR [34], ILS-VNS/VND [33], and TS-PR [35] the procedure can converge to close to optimal solutions even for very large-scale instances with homogeneous/heterogeneous fleet running with up to 10,000 employees and up to 500 vehicles. These solutions can be obtained in few minutes (**Figure 12**).

3.3 Spatial DSS to staff collection and dispatching

Graphvs™ developed a spatial DSS solution using its SisRot Full® latest version to help staff collection and dispatching location-routing and scheduling. In **Figure 14** we have the description of the architecture of this DSS framework.

The system is also implemented in a cloud unity (VPS) where its components can share information for decision analysis. The major system is composed in two parts, the first is the spatial decision support system (SDSS) SisRot Full where exists a Graph-Based Modeling System (GBMS) that operates with many tools over the street network, and the second is a set of tools to manage information from staff (their address, place of work, shift, etc.), garages and places of work, available fleet and vehicles' tracking, and staff transportation monitoring.

In (**Figure 13**), the SDSS of the SisRot Full® has the same previous characteristics available for the wholesales distribution and includes the tools and algorithms necessary to support decision to locate candidate closest bus-stops to staff address, to select the ones of minimum cost of location and edit solutions, to perform routes to collect or dispatch staff. These algorithms perform the preliminary solutions that can be edited

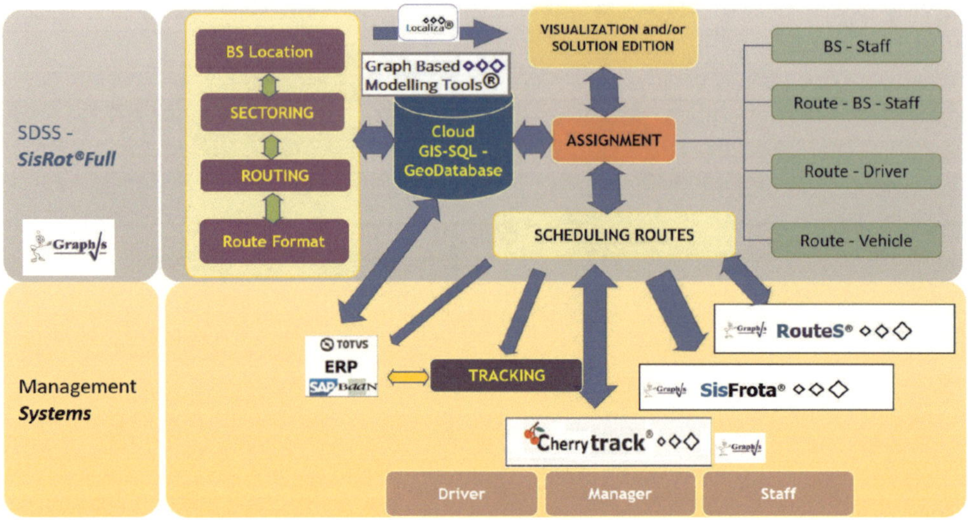

Figure 13.
Spatial decision support system for staff collection and dispatching logistic planning.

using productions available to move customers between bus-stops, routes, reassign routes to new garages or final stops, create or delete routes, and many other. The final solution can be sent to be viewed in different map tools such as Google Maps™ and Here Maps™ by using the software Routes®, to better evaluate the paths to bus-stops selected and routes accordingly to the hour to go from/to bus-stop/work. The final routes solution can be also assigned to the driver and can be scheduled accordingly the shift of the days of covering the staff.

External software for managing the maintenance of vehicles can be included such as SisFrota®, and managing staff such as ERPs from Baan, Totvs, SAP, and many other. Graphvs™ most recently introduced a new tool to follow staff while in the field, it is called CherryTrack®. In this tool it is possible to follow any staff while going/back to/from work, and it can give all the information about his/her way to/back work/home as position of the bus and approximate time to be at his bus-stop or to leave.

4. Health system

The use of spatial DSS for health system is still not evident in the health management environment. DSS is in general tool for planning, but at least in Brazil, this concept is avoided managers think like: "the problems are running and have to be solved just with the available resources in the meanwhile."

After understanding the particularities of the tactical and operational aspects of the organization of each macro problem, it is possible to evaluate adequate techniques to better approach the optimized organization. It is in fact difficult to achieve, but never surrender.

Following we discuss two important problems of main concern worldwide: Arbovirusis prevention, control, and combat using dengue as example; and emergency medical systems, as they are run in Brazil, and as we here propose new developments to them.

4.1 Dengue disease control and combat

Dengue is still a very dangerous tropical disease provoked by arbovirus on infected mosquitoes *Aedes aegypti*. It hurts millions of people around the world annually. The mosquito is now responsible to carry out virus not only for dengue, but chikungunya, yellow-fever, and zica, other dangerous diseases that hurt and kill in the same tropical areas many human beings. The capture and extermination of the mosquito are the major effort necessary by health municipality secretaries to avoid these diseases in Brazilian urban areas.

Negreiros et al. [16] explain a number of clustering applications to hold the preparation of different kind of problems to define the logistic of dengue disease control and combat in tropical areas. The most important work in this process is made by sanitary agents, actually called endemic agents.

The problem to manage endemic agents for covering restricted areas to control focal and strategic points can be solved by the same generalized capacitated clustering model, even for blocks or unitary real states that are spread in daily urban areas of control.

From the gCCCP problem, it can be only inserted the canalized constraints (68) to define areas of coverage limited to the agents minimum and maximum capacity to visit real states in an epidemiological circle.

$$Q_{min} \leq \sum_{i \in I} q_i \; y_{ij} \leq Q_{max}, \; j \in J. \tag{68}$$

4.2 A DSS to this problem (Webdengue: manager)

A framework of computational systems Webdengue was designed to treat the problem of endemic agents coverage in its many forms and populate with solutions to better assign and optimize resources. The framework counts with a manager system that is incorporated with the capacitated clustering DSS tool (calculate and edit solutions) and the assignment tool to send to agents' and supervisors' mobile systems the

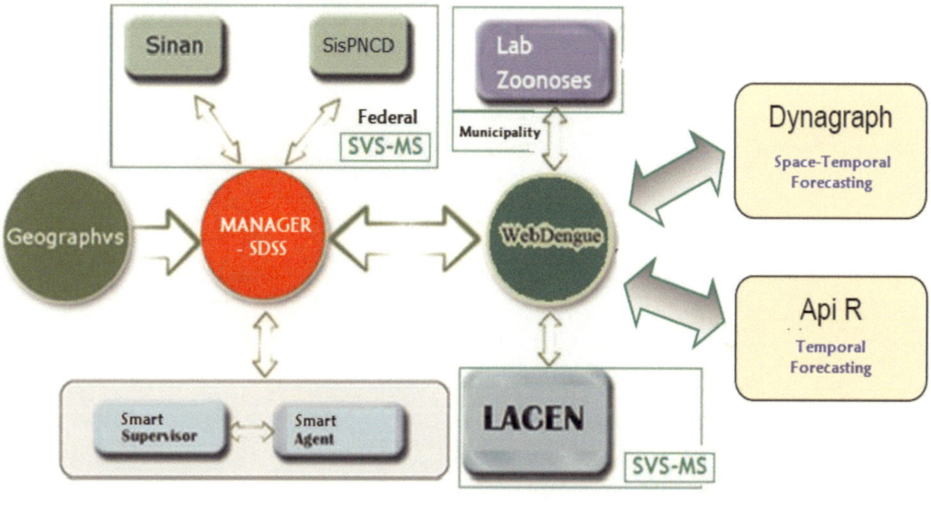

Figure 14.
New Webdengue computational framework.

routes between real states to cover. In **Figure 14** it can be seen the new structure of the framework, since its modification performed in 2021.

The framework contains a system to process the information from the field that comes from the mobile units and also that could be generated locally by the manager about the status of visits and people affected by the arbovirusis, **Figure 15a**. Tools to evaluate the territorial spread of the mosquito and the linkages with the human cases were developed to permit planning to better prevent the outcome of the disease, **Figure 15b**.

Most recently, Negreiros et al. [36] reported that it was included space-temporal clustering tools to track the disease in the affected territories by using a new dynamic-graph tool called Dynagraph, **Figure 16a**. Dynagraph is a space-temporal graph editor developed by Calixto [37] and later improved by [38], which works as a data warehouse to be adjusted to as many space-temporal applications as possible such as epidemiological outbreaks of diseases, grid network expansion, aircraft transportation network expansion, crime expansion, and so forth.

With this tool it is possible to see the changes between epidemiological weeks of the territorial occurrences of the *Aedes* and human cases, and apply clustering methods to forecast in the territory where will be in the next epidemiological week the

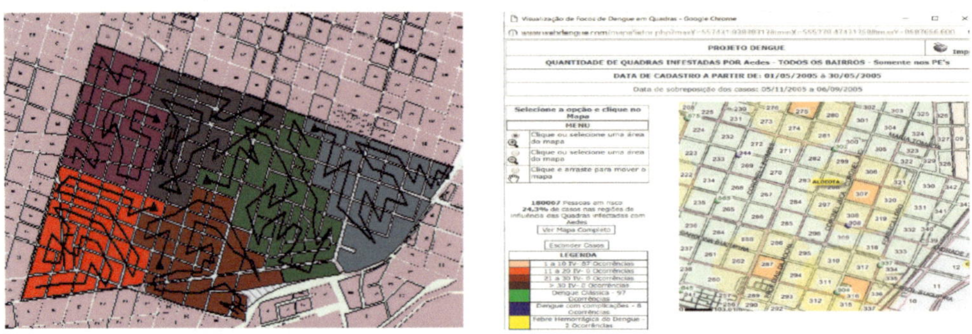

(a) (b)

Figure 15.
Planning with Manager and visualizing focus and cases in city blocks with Webdengue. (a) Sectoring and routing epidemiological agents between city blocks. (b) Focus X Human cases information cross.

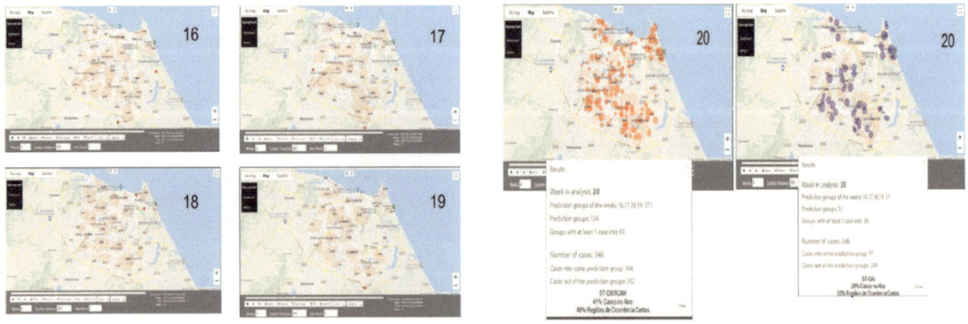

(a) (b)

Figure 16.
Using Dynagraph for space-temporal human cases prediction. (a) Evolution of cases between sequential epidemiological weeks. (b) Forecasting cases using st-clustering techniques.

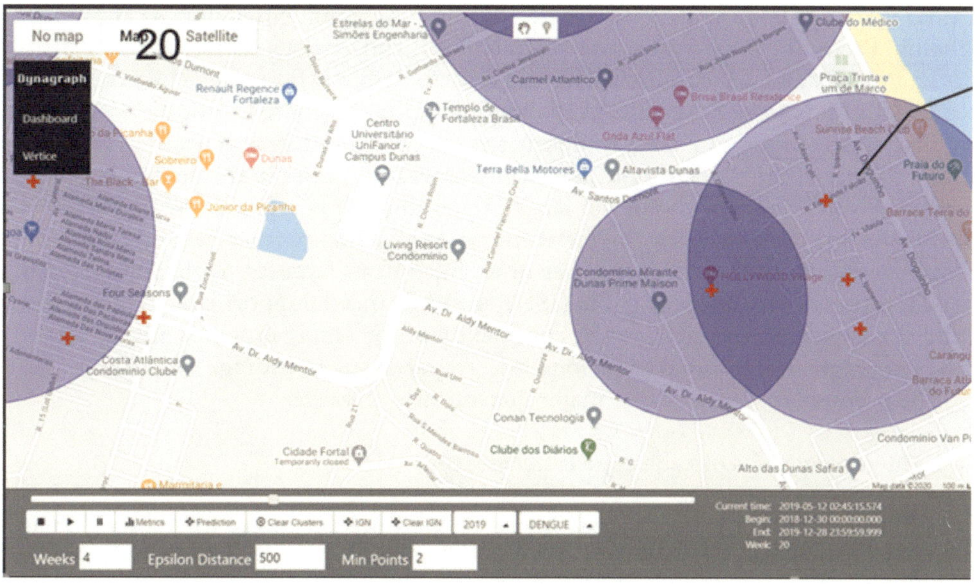

Figure 17.
Predictable areas of dengue occurrence.

places with cases of dengue, i.e., **Figure 16a** and **b**. In **Figure 17** we can see target circles where in this territory it is predictable to have human cases of dengue in epidemiological week 20.

4.3 Urgent ambulance system

SAMU (Serviço de Atendimento Móvel de Urgência, or *Emergency Mobile Care Service*, the same as EMS – *Emergency Medical System*) is a gratuity pre-hospitaller system operated by the municipality 24/7 that is available in major Brazilian cities and metropolitan areas, [39–41]. The logistic system is composed of four structures: calling reception, urgency identification and dispatching, urgent mobile care, and emergency hospital and clinics.

The calling reception and dispatching system (Central of Regulation) is composed by specialized people and doctors that receives urgent calls, filter and dispatch the real urgent cases to be serviced by a specialized unit and assigns the case to an available and appropriate health unity. The urgent mobile care system is composed of a huge structure of specialized ambulances that are equipped with appropriate care materials, instruments, and human resources (doctors, nursers, paramedics, etc.) to deal with the specific levels of urgent treatment needed to maintain the survival ability of the patients. Once the patient injury is confirmed in the field by the local health agents in operation, the calling operating system may renegotiate the better place for the service. The ambulance goes directly to the assigned health unity, and after safety delivery of the patient it is free to attend a new call.

The emergency service is regulated by international standards defined by World Health Organization (WHO). The response time, i.e., the elapsed time between notification of an occurrence and the ambulance arrival at the scene, is about 8 min, Pons & Markovchick [42]. This is the main performance measurement (challenge) of the

SAMU system. The Brazilian studies reveal that in medium cities such as Niterói/RJ, Ribeirão Preto/SP, Ouro Preto, and Mariana/MG, and capitals such as Belo Horizonte/MG and Rio de Janeiro/RJ, the response time is far to be achieved in capitals, but it is close to normal in the medium cities evaluated, [40, 43–45].

The logistic problem here is to define the appropriate fleet and health team to attend people in the diversity of the circumstances that exist in the emergency health context. From other worldwide similar systems and by experience, the players (municipality) already know the different types of infrastructure needed to attend people. In Brazil, there are three types of mobile infrastructures: Basic Support Units (BSU), Intermediate Support Units (ISU), and Advanced Support Units (ASU). These units are composed accordingly the complexity of the emergency it is able to attend, and it is assigned preferable to the one it is prepared for attending.

The problem resides in three major questions to answer:

1. How many units of each type is necessary?

2. Where (or how) is the (shape) region or locations they have to be based during their daily service time?

3. What hospital do they have to be assigned to achieve the international standards?

Once the system works with great uncertainty on where the occurrences will be in the space-time, the logistic problem is solved by adjusting a M/M/m Queueing Hypercube model that considers atomic regions and their probability of emergency occurrence for each type of service, based on historical data, [40, 41, 46, 47]. By using this model with the real dataset, the service units are placed along the atomic regions, and the simulation is performed to first evaluate the equilibrium of the service in a day of work, once the service times are satisfied, it is then observed the effect of removing units from different regions, [41].

For major cities it may be relevant to consider several bases in the infrastructure. These are places where the teams meet and the regulation center is installed, as the inventory of health care material and the ambulances. In less concentrated cities, the bases can be placed in hospitals when they are located in well-separated regions of the city. But there are cities, as shown by Nogueira Jr. et al. [48], that evaluated optimization models in conjunction with simulation, where they verified that their deterministic approach could minimize the time of service by reconfiguring the fleet, reducing active bases and ambulance relocation. This research brought the essence of the new model we are proposing here.

As it can be seen, the models proposed until now do not take care of reconfiguring the coverage areas as the cities grows or as the number of valid calls come. There is no capability of reaction, and no possibility of dynamic infrastructure revision while the service is on the way. The congestion of the system that works on the normal conditions has been evaluated, but the breakthrough of the actual infrastructure is not considered. Dynamic location optimization models may be introduced for this context. These models may concern to the stability of the system even if sporadic congestion occur in the system.

In the CCP context, there are two possible problem approaches to be used: Dynamic Generalized Multi-Capacitated Clustering Problem (DgMCCP) and a Dynamic Generalized Multi-Capacitated Centerd Clustering Problem (DgMCCCP). In the first model the bases are important to be selected for the logistic context, in the second model there is no base in fact needed to be selected, they are the hospitals that operate in the region defined by the cluster at the end. Below we consider the DgMCCP.

Consider that a base can be either a hospital or a place without hospitaller infrastructure but an infrastructure to keep medicine and other medical inventory and garage to maintain the ambulances. Consider each candidate to be a base (median) has a certain capacity for each type of service. Finally consider that a client must have only one type of service. The following notation for the generalized Multi-Capacitated Clustering Problem (gMCPP) can be placed:

I: The set of patients,

J: The set of candidate bases (medians),

K: The set of types of services.

H: The set of hospitals (with emergency units).

n: Cardinality of the set I, i. e., number of patients.

m: Cardinality of the set J, i. e., number of bases (medians).

CI_j: cost to install a base j as center.

CA^k: cost to allocate an ambulance of type k.

CTP^k: transportation cost of an ambulance of type k.

t_{ij}^{hk}: Total time to attend the patient i from the (candidate) base j with service k to hospital h.

q_i^k: Demand of patient i with service k (Unitary).

Q_j^k: The capacity of the (candidate) base j for the type of service k.

QA_j^k: The maximum number of ambulances based on j of service k.

NA^k: The number of ambulances of service type k.

$$x_{ij}^{hk} : \begin{cases} 1, \text{if patient } i \text{ with service } k \text{ is assigned from base } j \text{ to hospital } h. \\ 0, \text{otherwise} \end{cases}$$

$$y_j : \begin{cases} 1, \text{if base } j \text{ is used.} \\ 0, \text{otherwise} \end{cases}$$

$$\begin{aligned}(\text{gMCCP})\text{Minimize } & \sum_{j\in J} CI_j \; y_j + \sum_{k\in K} CA_k \; NA^k \\ & + \sum_{i\in I}\sum_{j\in J}\sum_{h\in H}\sum_{k\in K} CTP_k \; t_{ij}^{hk} \; x_{ij}^{hk},\end{aligned} \tag{69}$$

$$\text{such that : } \sum_{i\in I}\sum_{k\in K} x_{ij}^{hk} \geq y_j, j\in J, h\in H, \tag{70}$$

$$\sum_{j\in J}\sum_{h\in H}\sum_{k\in K} x_{ij}^{hk} = 1, i\in I, \tag{71}$$

$$\sum_{i\in I}\sum_{h\in H} q_i^k \; x_{ij}^{hk} \leq Q_j^k \; y_j, j\in J, k\in K, \tag{72}$$

$$\sum_{k\in K}\sum_{h\in H} x_{ij}^{hk} - y_j \leq 0, i\in I, \; j\in J, \tag{73}$$

$$\sum_{k\in K}\sum_{h\in H} x_{ij}^{hk} = NA_k, \; i\in I, \; j\in J, \tag{74}$$

$$\sum_{j \in J} QA_j^k \leq NA^k, \quad k \in K, \tag{75}$$

$$x_{ij}^{hk}, y_j \in \{0,1\}, NA_k \in \mathbf{N}, \quad i \in I, \quad j \in J, \quad k \in K, \quad h \in H. \tag{76}$$

The gMCCP considers as objective (69) the minimization of the fixed cost to use the structure of a base j and ambulances of types k, and the variable cost to attend all patients to be serviced. In constraint (70) it is warranted the assignment of a base j and than to the hospital h if there is at least one service to it. In constraint (71) it is guaranteed that the service of a patient is assigned to at least a base hospital. In constraint (72) the capacity of a base for any type of service stays at its limits. Constraint (73) reinforces the existence of a base if there is a service to it. In constraint (74) it is guaranteed availability of the number of ambulances to attend the patients. Constraint (75) guarantees the number of ambulances per type is limited. Constraint (76) defines the decision variables.

The dynamic nature of a clustering problem can be modeled as a linear function of time as in the dynamic location problem, [49]. The uncertainty parameters of the model are most often described through scenarios. A scenario is a possible realization of uncertain parameter values, [50–52]. In our approach, we consider the set of fixed facilities along the planning horizon, with the adaptability to the service (growth or decline) accordingly to the conditions of each scenario. In this approach, the concept of regret function is used, which searches to minimize the impact of bad choices of a set of facilities among the different scenarios of the planning horizon.

Consider that a scenario is composed by the patients of a single shift or workday. A set of scenarios S is composed by a number of shifts or workdays that can be extracted from the data set of services realized during the period of analysis. Consider as a regret function to locate the bases, the function that minimizes the maximum difference of cost to cover all customers per scenario using the best configuration and actual scenario configuration. The new model can be placed as:

S: The set of scenarios.

r: is the final cost of the optimal regret.

$z^\star(x, y, NA)$: is the cost of the optimal solution between scenarios.

$z^s(x^s, y^s, NA^s)$: is the cost of the solution of scenario s.

$g_{ij}^{hks} = x_{ij}^{hks} = x^s$:

$$g_{ij}^{hks} = x_{ij}^{hks} = x^s : \begin{cases} 1, \text{if patient } i \text{ with service } k \text{ is assigned to base } j \text{ to} \\ \text{hospital } h \text{ in scenario } s \\ 0, \text{otherwise} \end{cases}$$

$$y_j^s = \begin{cases} 1, \text{if base } j \text{ is used in scenario } s \\ 0, \text{otherwise} \end{cases}$$

$$\text{(DgMCCP) Minimize} \quad r, \tag{77}$$

$$\text{such that}: \quad |z^\star(x, y, NA) - z^s(x^s, y^s, NA^s)| \leq r, \quad s \in S, \tag{78}$$

$$x_{ij}^{hk} \geq g_{ij}^{hks}, \quad i \in I, \quad j \in J, \quad k \in K, \quad s \in S, \tag{79}$$

$$y_j \geq y_j^s, \quad j \in J, \quad s \in S, \tag{80}$$

$$r, z^{\star}(x,y,NA), z^{s}(x^{s},y^{s}), NA, NA^{s} \in \mathbf{N}, \; s \in S, \tag{81}$$

$$x_{ij}^{hk}, y_j, g_{ij}^{hks}, y_j^{s} \in \{0,1\}, \; i \in I, \; j \in J, \; k \in K,$$
$$h \in H, \; s \in S. \tag{82}$$

Constraints (78) should be replaced by:

$$z^{\star}(x,y,NA) - z^{s}(x^{s},y^{s},NA^{s}) \le r$$

and

$$z^{s}(x^{s},y^{s},NA^{s}) - z^{\star}(x,y,NA) \le r.$$

The gMCCP considers as objective (77) the minimization of the regret function. Constraint (78) reports the maximum regret. Constraint (80) guarantees the coverage of the optimal solution in the optimal solution of the scenarios. The constraint (81) guarantees the bases (medians) used in the optimal solution can also be used in the optimal solution of the scenario. Constraints (81) and (82) are the decision variables dominion.

This dynamic model (DgMCCP) can be considered as a better and stimulating challenge for the SAMU problem, once it considers the real data as inputs. The simulation step can also be used to evaluate the precision of the model and its response. The model in itself is in NP-Hard complexity, as the versions proposed by [48, 53]. Important to notice that the bases are not known in advance (although there are given candidates), and it is an output of the model between scenarios, as the number of ambulances per type.

5. Waste collection

Waste collection actually is considered as an activity that develops and operates different logistic systems to give to solid waste management (SWM) that optimized plans that guarantee the best performance and minimal costs involved. SWM is composed by services such as: domiciliar and selective waste collection, container collection, rubble collection, health residues collection, human and mechanical street sweeping, and many others.

We discuss four problems related to waste collection: domiciliar, selective, street sweeping and collection with hill climbing. In these problems, most of them are traditional, and relevant literature considers them. The eye on arc sectoring routing is the most advanced because it changes the way the waste collection is performed in Brazil and worldwide. We used CCP models to obtain solutions to the first step of these problems, the sectoring, but we also show the final step by using our arc routing algorithms.

5.1 Domiciliary

The domiciliary refuse collection is a regular activity worldwide, spending billions of dollars monthly in a reverse logistic system where the garbage is collected on streets of residential and/or mixed with commercial areas and discharged at disposal sites like transfer stations and/or landfills using special type of vehicles (i.e., compactors, load packers).

In Brazil, most collection is held by contractors or concessionaires that do their service to municipalities. They divide cities into zones of collections that are covered regularly in specified periods like (diary, alternately or accordingly to the demand). The zones are divided into sectors, areas that are dimension-ed to be covered by vehicles within local workload per day. Each sector of a zone is also covered in several trips, once the vehicle is fulfilled it goes to discharge and comes back to a new loading trip. This circle repeats until all garbage is collected in the sector.

To guarantee that the fleet of vehicles covers all the sectors in the defined workload per day, it is necessary to generate good estimations of the garbage load behavior along a reasonable period. Once it is possible to have good forecasts for garbage generation for a long period, it can also be drawn regions in the sectors to traverse with one vehicle that will guarantee the coverage of the sector for all trips during the workload. This is called the Arc Sector-Routing problem (ASRP), a recent problem proposed by the Mourão et al. [54], which is being studied by a number of researchers, [55–60].

Most recently, Negreiros et al. [56] proposed three methodologies based on heterogeneous capacitated centerd clustering problem (HCCP) to produce automatic near-optimal solutions to the heterogeneous arc-sectoring routing problem (HASRP). The first method builds solutions by sectoring-clustering-routing (SCR) where sectoring is the process of defining the great areas to be covered in a workday using different types of vehicles (different capacities) and then internally the coverage of the trips is defined in capacitated clusters limited to the maximum load of the vehicle and then the routing phase calls the rural postman routes to perform the trip over the clusters formed; the second process indicated was sectoring-routing-clustering and routing (SRC&R) where it defines sectors and the routes following that giant routes are build into the sectors, after the route is divided into circuits by using an appropriate method, i.e., as the one proposed by [61]; and finally the sector-routing-clustering-routing (SRCR) where it performs the sectoring using the HCCCP and after it generates a giant rural postman tour that is divided into trips of known capacity by using [61] and rebuilt the rural postman tours for each final trip.

The Heterogeneous Capacitated Centerd Clustering Problem with Fixed Cost (HCCCP-FC) here used for the sectoring/clustering phase can be defined as:

$L = \{1, 2\}$: is the dimension of the Euclidean space.

I: is the set of arcs.

S: the set of sector centers. We consider $|I| > |J|$.

V: the set of different vehicles to be used.

$a_i \in \mathbf{R}^{|L|}, i \in I$.

q_i: the garbage offer of an arc $i \in I$.

Q_s^v: the maximum garbage offer of a sector $s \in S$, using vehicle v.

c_v:, $v \in V$ is the cost of using a vehicle of capacity v.

VC: Variable cost.

$\overline{x}_s \in \mathbf{R}^{|K|}, s \in S$ represents the center of *sector* $s \in S$.

n_s an integer, is the number of arcs in *sector* $s \in S$.

$$y_{is} = \begin{cases} 1, & \text{if } i \text{ is an arc in sector } j \\ 0, & \text{otherwise.} \end{cases}$$

$$w_s = \begin{cases} 1, & \text{if sector } s \text{ with vehicle type } v \text{ is used} \\ 0, & \text{otherwise.} \end{cases}$$

The HCCCP-FC can be formulated as follows:

$$(\text{HCCCP} - \text{FC})\text{Minimize}\sum_{s\in S}\sum_{v\in V} c_v\ w_v^s + VC\ , \tag{83}$$

$$\text{such that}: \sum_{v\in V}\sum_{i\in I}\sum_{s\in S} \|a_i - \overline{x}_s\|\ y_{is}^v = VC\ , \tag{84}$$

$$\sum_{v\in V}\sum_{s\in S} y_{is}^v = 1, \quad i\in I, \tag{85}$$

$$\sum_{v\in V} w_s^v = 1, \quad s\in S, \tag{86}$$

$$\sum_{i\in I}\sum_{s\in S} y_{is}^v - w_s^v \geq 0, \quad v\in V, \tag{87}$$

$$\sum_{i\in I}\sum_{v\in V} y_{is}^v \leq n_s, \quad s\in S, \tag{88}$$

$$\sum_{i\in I}\sum_{v\in V} a_i\ y_{is}^v \leq n_s\ \overline{x}_s, \quad s\in S, \tag{89}$$

$$\sum_{i\in I} q_i\ y_{is}^v \leq Q_s^v\ w_s^v, \quad s\in S,\ v\in V, \tag{90}$$

$$\overline{x}_s \in \mathbf{R}^{|K|}, n_s \in \mathbf{N}^*, s\in S, \tag{91}$$

$$y_{ij}^v \in \{0,1\},\ w_s^v \in \{0,1\}, \quad i\in I, s\in S,\ v\in V\ . \tag{92}$$

The same transformations that were performed on gPCCCP to transform it into gPCCCP' can be performed on HCCCP-FC.

The objective function (83) minimizes the total fixed cost of the fleet used and the total Euclidean distance between *sector* centers and their assigned arcs. Constraint (84) refers to the total variable cost. The constraint (85) assigns one arc to just one *sector*. The constraint (86) considers the number of arcs per *sector*. The constraint (87) assigns one vehicle to just one *sector*. The constraint (88) considers once a vehicle is used to service an arc it is assigned to a *sector*. The constraint (89) defines the geometric center of the *sector*. The constraint (90) limits the assigned arcs to the maximum capacity of the *sector s*. The constraints (91) and (92) refer to the decision variables of the problem.

In **Figure 18** we can see a sectoring procedure operating to divide a zone of domiciliary waste collection selecting from a set of (19 t, 15 t, and 10 t), the model returns two different vehicles (19 t and 10 t) forming three circuits for 3x39t and 1x27t sectors for the Aldeota and Meireles boroughs in Fortaleza/CE.

5.2 Selective

The selective garbage collection is an alternative service of waste collection that is being largely explored in Brazil. More than 5% of the garbage can be recycled to be re-manufactured or to the industry (paper, glass, plastic, aluminum, iron, hazardous materials – batteries, oil, etc.). The residential garbage is most organic as can be seen in the literature about the residential garbage gravimetry, [62, 63]. Actually the cities occupation of commercial and domiciliary unities is mostly fuzzy, opening spaces for different kind of selective waste collection.

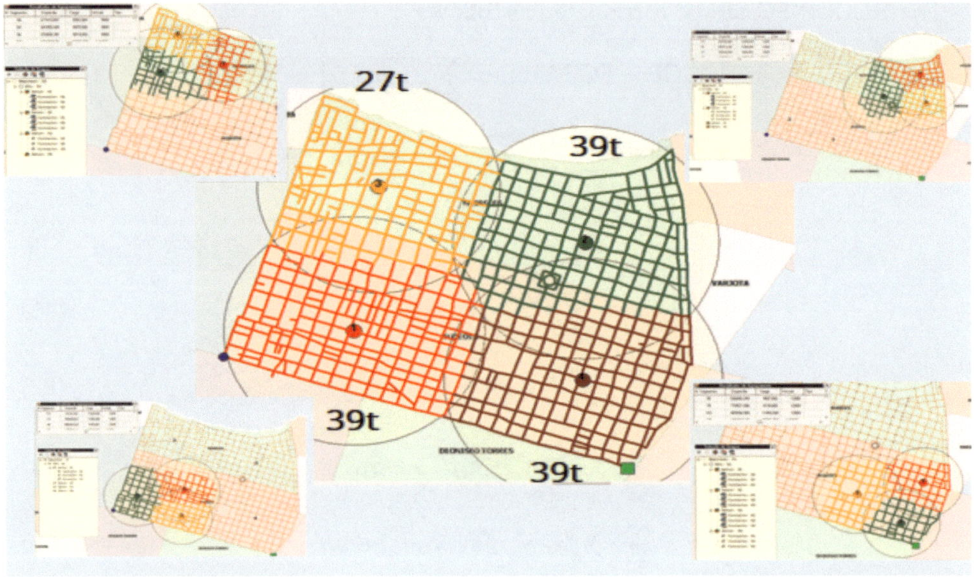

Figure 18.
Heterogeneous sectoring and circuiting for domiciliary waste collection in Aldeota and Meireles - Fortaleza/CE.

There are many types of selective waste collection, but three of them are most used in Brazil and many other countries around the world: selective household collection using compactors, selective household collection using vehicles equipped with "jail" units, and selective collection by agents and big compactors with roll-on-roll of equipment to take big boxes of separated garbage (plastic, cardboard, glass, and metal).

In the selective household collection using compactors, the work is done in a specific day of the week where residents keep their selected dry garbage at door. A compactor with huge volumetric capacity is assigned to the region to collect all the material taking into consideration that the volumes can be stored in the compactor unit for one or two trips during a workload. A driver and two garis compose the team for each vehicle in general. The work is performed by concessionaires and/or municipalities, and it is directly sent to triage centers that are runner or organized by associations of garis, [64–66].

The selective household collection using vehicles equipped with "jail" does the same work, although it is not possible to compact the garbage. It takes two people to arrange the volumes into smaller ones in the vehicle of 28–31m^3 (5 m X 2.9 m X 2 m). It is mostly performed by small associations and cooperatives recognized by the municipalities. It is worth to say that the Brazilian law 12,305/2010 turns the industry responsible for the garbage resulting from the discharge of its final consumer. There are specialized organizations that perform all the preparation of these associations from projects of industrial triage plants to business management, using the resources from industry to obtain the ecological certification (residue free, emission free) in carbon-free credits. The infractors have to pay heavy fines.

The selective household collection using agents is the more sustainable system that guarantees to the "cleaning agent" job placement and a special place in the society. This system is composed by two stages, the first is performed by the collection agent, which does the selective collection using manual vehicles or motorcycles equipped with small

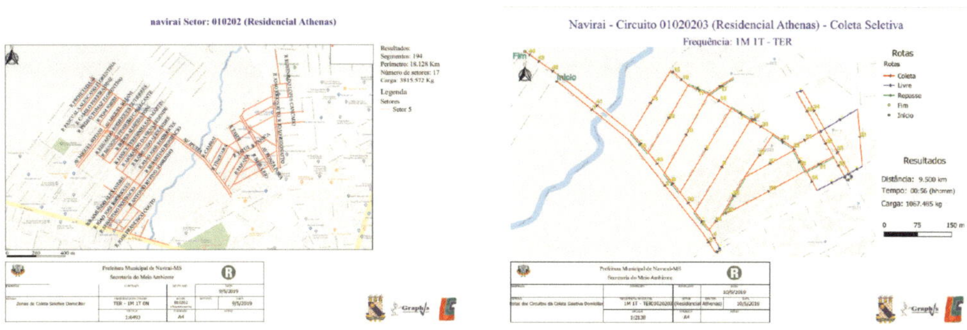

(a) (b)

Figure 19.
Sectors and a circuit of the selective domiciliary waste collection for Navirai/MS. (a) A sector of the selective domiciliary waste collection for Navirai/MS. (b) A route in a circuit of the selective waste collection for Navirai/MS.

"jails" ($0.5m^3$) to store the different type of selective garbage. They collect in small volumes and cover small to medium areas (10–20 blocks). They also do street-sweeping in the days they do not perform selective collection. The garbage goes to special boxes, which are collected by big vehicles equipped with compartments and roll-on-roll-off units that carry the garbage to triage centers. The result of this process pays for the salary of the agents and supports the concessionaires and/or associations, with double return to the municipality.

The models used to plan the service of the first two types of processes can be the same as the model proposed for domiciliary waste collection. In **Figure 19a** it is seen a sector of service projected for the city of Naviraí-/MS (Brazil) and the route for a circuit of this sector, **Figure 19b**. This solution was performed by GRAPHVS for the Brazilian organization Recicleiros to selective waste collection using SisRot LIX®, [67, 68].

The model to plan service of the third process is composed by two phases of clustering, in the first it is drawn to the areas covered by the agents. It is the gCCCP with homogeneous capacity per agent. The second step is to define the routes of the big vehicles and define the location of the places where the sets of boxes will be in the street network on the intersection of clusters defined for the agents. **Figure 20** shows the solution produced by another system SisRot LIX-SELECTIVE® considered for the third form of selective collection, using agents and bigger vehicles. Also, it can be seen the two steps of the logistic system (**Figure 20**).

5.3 Street sweeping

Public sweeping is a very difficult process to organize because its planning depends on the flow of people, vehicles, and afforestation in public places. To meet this demand, the city is divided into regions (or sub-municipalities) with defined sweeping frequencies and sectors for groups of sweepers. Circuits are places where each team of sweepers must perform the sweeping task. This service is inspected in sector groups. This service is costly; it demands over 40% of the total cost related to the non-scaled services of the São Paulo city, approximately 250 million USD/yr., [69].

Sweeping is performed in many manners: manual, mechanized, on sidewalks, in squares, in public installations, etc.; however, this study focus on manual and

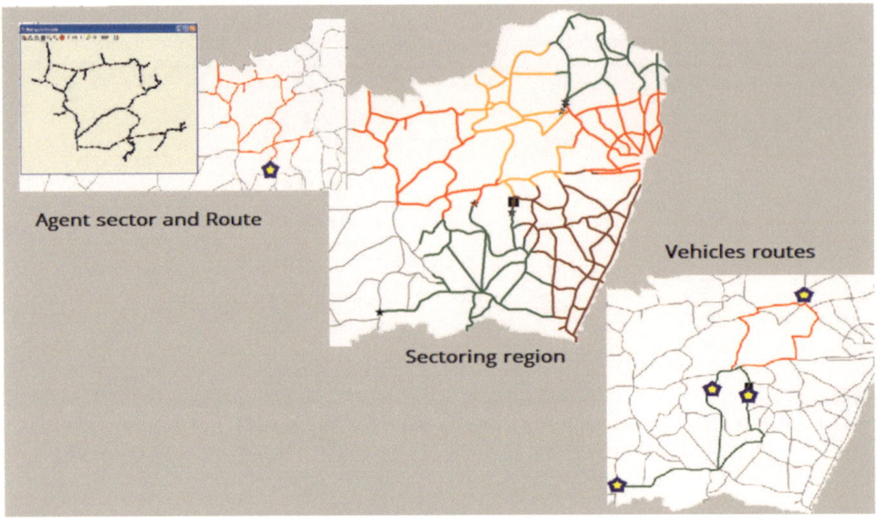

Figure 20.
Logistic system for cleaning agent in fixed selective collection sectors.

mechanical sweeping plans. In the manual sweeping of São Paulo/SP the major Brazilian city, two street sweepers work together on the same gutter side (1.2 m within the sidewalk line), using a "lutocar," typical removal equipment, and appropriate clothing. Each team travels an average of 1.228 km/h sweeping and walking at a speed between 2.5 and 3.0 km/h. At this rate, a route of 3.8–5.6 km per working day (7 h 20 min) can be accomplished.

5.3.1 Manual sweeping

The manual sweeping in the city of São Paulo is paid by considering the street axis perimeter. This approach indicates that a route is taken by a team of two sweepers with a "lutocar" working together that do for the sector perimeter at the half-time distance of the team workload (7 h 20 min). The sweep begins from a designated starting point and goes to an end-point passing through a set of street segments, in the minimal walking path cost, **Figure 21a**. For the streets that must be swept in the first route passage, the sweepers perform the service in the gutters on one side of the road. Upon returning, they perform the service in the gutters on the other side, **Figure 21b**.

Another way to perform the same work is by dividing the sweepers to sweep each gutters street side simultaneously, **Figure 22a**. For this case, there is only one open route passage, and the sweepers can perform the same service by walking much less but each one needs to operate with a "lutocar". **Figure 22b** displays two examples where an open sweep route (right) by axis is 48.26% shorter than the path from the same sector coverage (left) and in the second the open route is 36.68% shorter than the path using both sides. This solution was proposed to the municipality of São Paulo to reduce labor fatigue and increase sweeping system coverage, [69].

As seen in **Figure 22**, the route designed for the sweeper covers the same set of public places. Once it was determined to include a fixed starting point to allow the sweepers could find a unique place to keep the "lutocar," we observed that the open RPP's minimum route can be successfully used. The team of sweepers begins from a

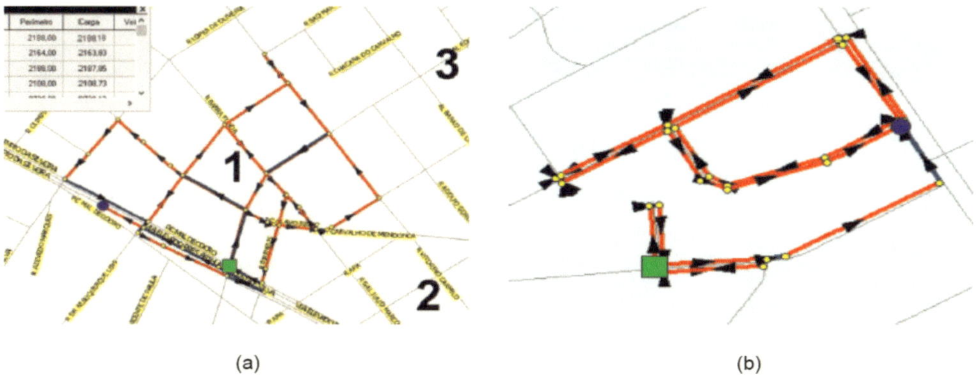

Figure 21.
Two different ways of human street sweeping. (a) Sweepers following the axis of the street. (b) Sweepers following gutter's sidewalks.

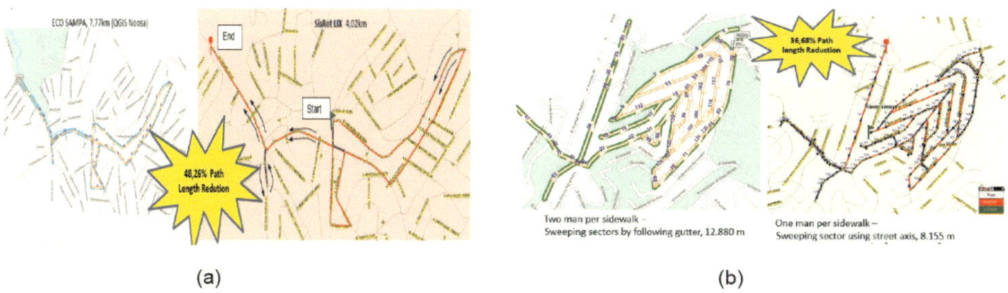

Figure 22.
Comparison between sweeping ways of human street sweeping. (a) Sector 01. (b) Sector 02.

support point and concludes their task at a specific point, minimizing the overall cost of the sweepers journey, which will then return to the support point.

An alternative sweeping service sector formation was also considered, where each sector exhibited a capacity that was limited by a maximum sweeping perimeter that cannot exceed half the work shift of each team assigned to it. Thus, the problem becomes an open arc sector-routing problem (OASRP) because a street sweeper team work in a sector, and any team may not overcome their workload, making it necessary to know the path of each team.

Sectors are projected for human street sweepers by the length of the cleaning path, that is most commonly approximated by the workload of a day. The sweep distances in meters are references to workload that can vary depending upon the slope and the pavement to be swept. In **Figure 23** it is projected a sector for three teams of two man, working at most for 2.8 km per day, in a defined sidewalk area for the sub-municipality of Campo Limpo/São Paulo-SP.

5.3.2 Mechanical sweeping

In mechanical street sweeping, the work is directly performed by the concessionaires on the most important roads, avenues, and marginals. The work is performed by large wash-sweepers equipped 1000–5000 L tanks of water, steel sweepers, and big vacuum cleaners. The sweeping velocity does not exceed 7–10 km/h, and two people work with the machine (driver and auxiliary). Road coverage occurs daily, but some

Figure 23.
Sectors for human sweeping in Campo Limpo-São Paulo/SP.

of them are covered three times per week at night. The frequencies are defined for each place according to their importance. In **Figure 25** it can be seen the region and the sector's solution proposed to the sub-municipality Campo Limpo/São Paulo for mechanical street sweeping.

The problem of mechanical sweeping is closely related to that of the human sweeper, although the network is mixed in this case, and the required links become required arcs when the traffic flow must be followed. We can solve the OSARP by using open routes. To design the open routes, we propose the use of the open MRPP (OMRPP), where we adapt SisRot LIX® to perform nearly or optimally for all routes. There are many ways to cross the street from the gutters on both sides, where the cleaning may be performed, but a single graph transformation can be performed to solve the OMRPP as a MRPP, [70–72].

By applying the methods developed for SisRot LIX®, we achieve superior results than those previously obtained by EcoSampa concessionary. The results demonstrate that it is possible to reduce the number of sectors and still reduce the total time spent on working by 49.20% and distance traveled by mechanical sweepers by 10.52% using their own fleet, **Figure 24**.

5.3.3 Domiciliary waste collection with hill climbing

The sectoring problem of domiciliary waste collection with hill climbing appeared during a project in the city of Petropolis/RJ – Brazil, to optimize domiciliary waste collection sectors and routes. The city of Petropolis/RJ is located along the Mantiqueira mounting range, where heights between 400 and 1000 m above the sea level in a bumpy relief are present, narrow and sloping streets are in general along the paving paths; these provoke great difficulty in up, down and collecting when compactors are used, because the weather is humid with frequently rain turning slippery and dangerous the collecting paths in heavy loads. Driving and waste collecting in such places are very difficult and dangerous. It is usual driving in reverse for more than 600 m while cars remain park along sidewalks.

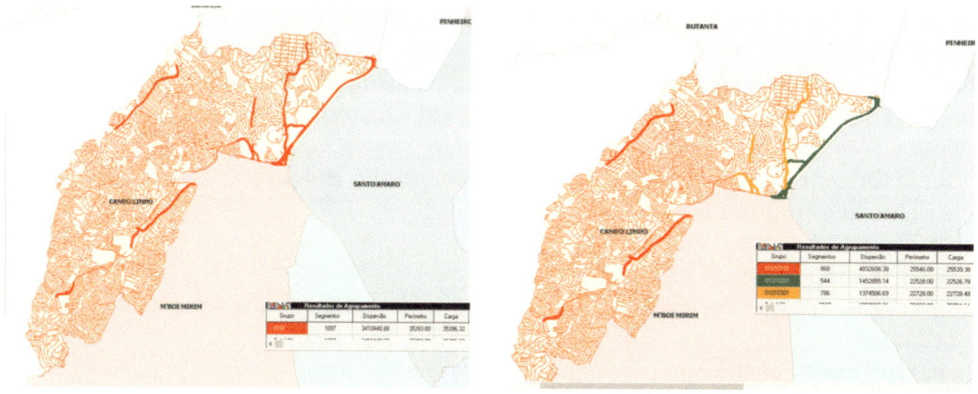

(a) (b)

Figure 24.
Sectoring mechanical street sweeping for Campo Limpo, São Paulo/SP. (a) Region of mechanical sweeping in Campo Limpo, São Paulo/SP. (b) Sectoring the region accordingly vehicle coverage constraints from Figure 24a.

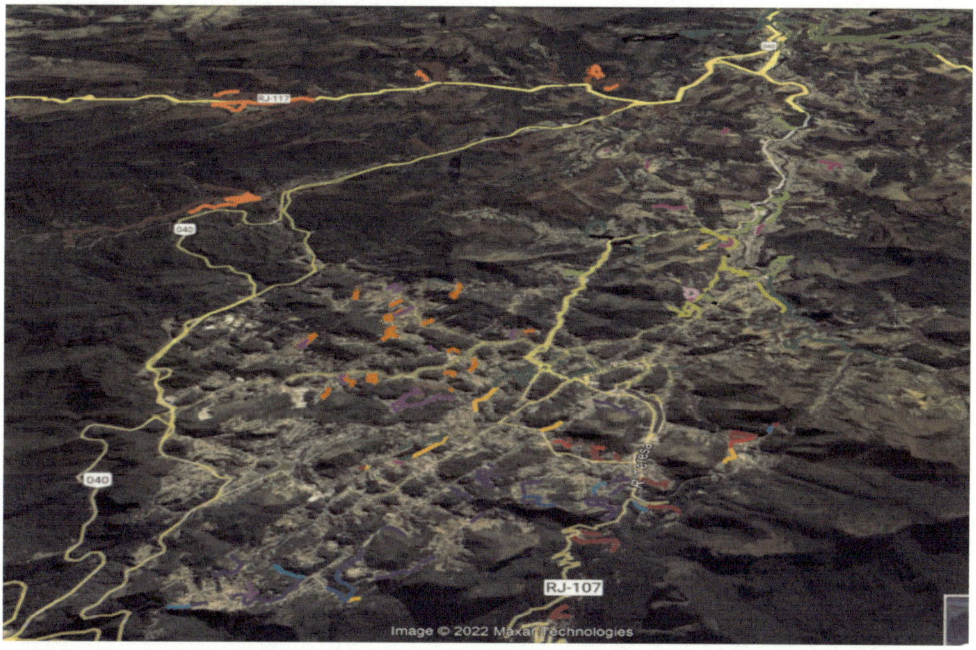

Figure 25.
Places where only "giricas" can cover.

Special compactors are used to do the waste collection, they are called "giricas." They have 3.5 t of load capacity at maximum, and they are smaller than the compactor used by the local operator. This vehicle is lighter also and faster. But it also suffers in high slopes and slippery paving paths. In **Figure 25** it can be seen a satellite photo from the area of Petropolis/RJ, and the places identified that are covered only by "giricas". In fact the paths with these characteristics are spread out along the city.

The contractor owns two types of vehicles that can cover the city, and only "giricas" are able to cover any path with moderate slopes. Their problem is to design capacitated clusters of sectors that cover a region with special service for required arcs considering the different capacities and quantities of their available vehicles. To make sectors using such heterogeneous and specialized fleet, the sectoring problem can be reduced to the Multi-Capacitated Centerd Clustering Problem (MCCCP), as follows:

L: Is the set of dimensions: L={Latitude, Longitude, Height}.

I: The set of required arcs.

J: The set of sectors.

K: The set of types of services.

$|L|$: Cardinality of the set L, i. e., number of dimensions of the geographical coordinates.

n: Cardinality of the set I, i. e., number of required arcs.

m: Cardinality of the set J, i. e., number of sectors.

$|K|$: Cardinality of the set K, i. e., number of types of service.

CS_j: cost to install a sector j.

CV^k: cost to allocate a vehicle of type k.

VV^k: variable cost per km to use vehicle of type k.

d_{ij}^k: distance between the center of a required arc i and the center of the sector j with service k.

a_i^l: Coordinate of the center of required arc i in dimension l.

q_i: Demand of required arc i.

Q_j^k: The capacity of sector j for the type of service k.

MS^k: The maximum number of sectors of service k.

NS_j^k: The number sectors j of service k.

NV^k: The number of vehicles of type k.

$\overline{x}_j^l$: the coordinate l of sector center j.

$$x_{ij}^k = \begin{cases} 1, \text{if required arc } i \text{ with service } k \text{ is assigned to sector } j \\ 0, \text{otherwise} \end{cases}$$

$$y_j = \begin{cases} 1, \text{if sector } j \text{ is used} \\ 0, \text{otherwise} \end{cases}$$

$$\text{(MCCCP)Minimise} \sum_{j \in J} CS_j \; y_j + \sum_{k \in K} CV_k \; NV^k + \sum_{i \in I} \sum_{j \in J} \sum_{k \in K} VV_{ij}^k \; d_{ij}^k \;, \tag{93}$$

$$\text{such that} : \sum_{l \in L} \left(a_i^l \; x_{ij}^k - \overline{x}_j^l \right)^2 \leq \left(d_{ij}^k \right)^2 \;, \quad d_{ij}^k \geq 0, \; i \in I, \; j \in J, \tag{94}$$

$$\sum_{i \in I} \sum_{k \in K} x_{ij}^k \geq y_j, \; j \in J, \tag{95}$$

$$\sum_{j \in J} \sum_{k \in K} x_{ij}^k = 1, \; i \in I, \tag{96}$$

$$\sum_{j \in J} \sum_{k \in K} q_i \; x_{ij}^k = q_i, i \in I, \tag{97}$$

$$\sum_{i \in I} q_i \; x_{ij}^k \leq Q_j^k \; y_j, \; j \in J, \; k \in K, \tag{98}$$

$$\sum_{k \in K} x_{ij}^{k} - y_j \leq 0, \quad i \in I, \; j \in J, \tag{99}$$

$$\sum_{i \in I} a_i^l \; x_{ij}^{k} = \overline{x}_j^l \; NS_j^k, \quad \forall j \in J, \; \forall l \in L, \tag{100}$$

$$\sum_{j \in J} NS_j^k = NV^k, \quad k \in K, \tag{101}$$

$$\sum_{j \in J} y_j \leq \sum_{k \in K} MS^k, \tag{102}$$

$$\sum_{j \in J} NS_j^k \leq MS^k, k \in K, \tag{103}$$

$$x_{ij}^{k}, y_j \in \{0, 1\}, \quad i \in I, j \in J, k \in K, \tag{104}$$

$$\overline{x}_j^l \in \mathbf{R}, J \in J, \quad l \in L, \tag{105}$$

$$NS_j^k, NV^k \in \mathbf{N}, \quad j \in J, k \in K \; . \tag{106}$$

Costraint (94) is second-order cone, which forms a convex set. The MCCCP considers in (93) the minimization of the fixed cost to use sector j and vehicles of type k, and the variable cost to service all required arcs. Constraint (94) defines the distance from a required arc to its center j for the service k. Constraint (95) warranties the assignment of sector center j to its required arcs with all services involved. In constraint (96) it is guaranteed that the service of a required arc will be assigned to a sector with a service. Constraint (97) guarantees the demand of a required arc is covered for any type of service. In constraint (98) the total waste offer of a sector j with service k does not overpass the capacity of the cluster j with service k. Constraint (99) guarantees the existence of a required arc to a sector if it is opened. Constraint (100) defines the center of the sector j with service k. Constraint (101) guarantees the number of sectors of service k is the number of vehicles of service k. Constraint (102) guarantees the number of opened sectors is at most the sum of number of maximum sectors available per service k. Constraint (103) guarantees the number of sectors of type k is at most the maximum number of sectors to service k. Constraints (104–106) define the decision variables.

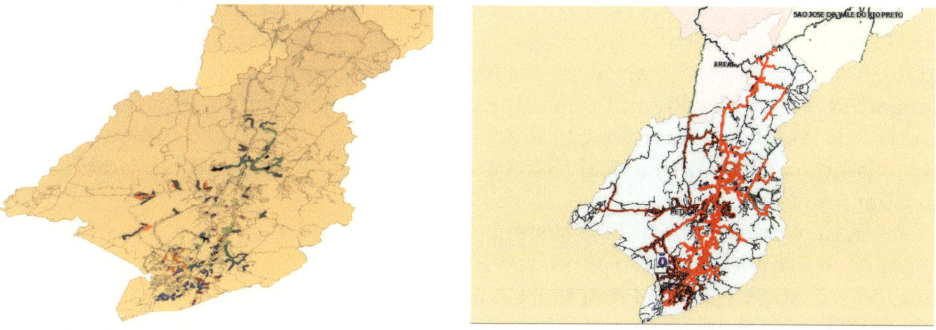

(a) (b)

Figure 26.
Region to be covered by domiciliar waste collection in Petrópolis/RJ. (a) Situation where only "giricas" can cover. (b) Region of coverage in Petrópolis/RJ.

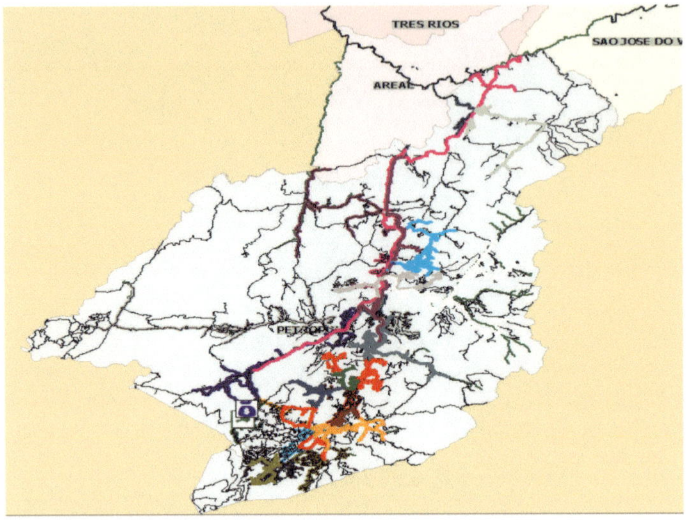

Figure 27.
Solution using "giricas" (3.5t) and compactors (8t) for Petrópolis/RJ.

The MCCCP problem is still NP-Hard as its version proposed by [73] and most recently deeply investigated by [74]. In **Figure 26a** this problem can be seen in a resolution process considering that the required arcs coverage must be considered and in **Figure 26b** it is seen the places where special required arcs must be covered only by "giricas."

The solution presented in **Figure 27** "giricas" may cover both types of required arcs, and they have to be placed in areas that maximize the coverage of the special required arcs, although they have also to cover non specials. The clustering process is supposed to warranty the separation between groups of service, but it does not guarantee that they will be non-overlapping clusters.

5.4 Spatial DSS to waste collection logistic planning

Graphvs™ developed for more than 30 years a spatial DSS to automatically solve sector-arc-routing problems for many different types of waste collection services. The spatial DSS is called SisRot LIX®, and it is being used with practical impressive results (up to 28% on savings) for applications in domiciliary waste collection (Campo Grande/MS, Petrópolis/RJ, and others), selective waste collection (Naviraí/MS, Casimiro de Abreu/RJ, Xanxerê/SP, Bom Jesus dos Perdões/SP, and others), street sweeping human and mechanical (São Paulo/SP), and most recently to Domiciliary Hill Climbing (Petrópolis/RJ).

The framework of spatial DSS systems is described by its architecture in the **Figure 28**. It is divided in two parts, the VPS-based tool and the mobile and management tools. In the VPS-based tool there is a georeferenced database linked to QGIS and an ESRI™-based component to read the maps and transform its information into a graph, by using an extended graph-based modeling system (E-GBMS), specially developed to waste collection. With the E-GBMS it is possible to include and edit the required network to design sectors and routes, and all the necessary tools to manipulate information about street loads and pathways characteristics. The system can solve

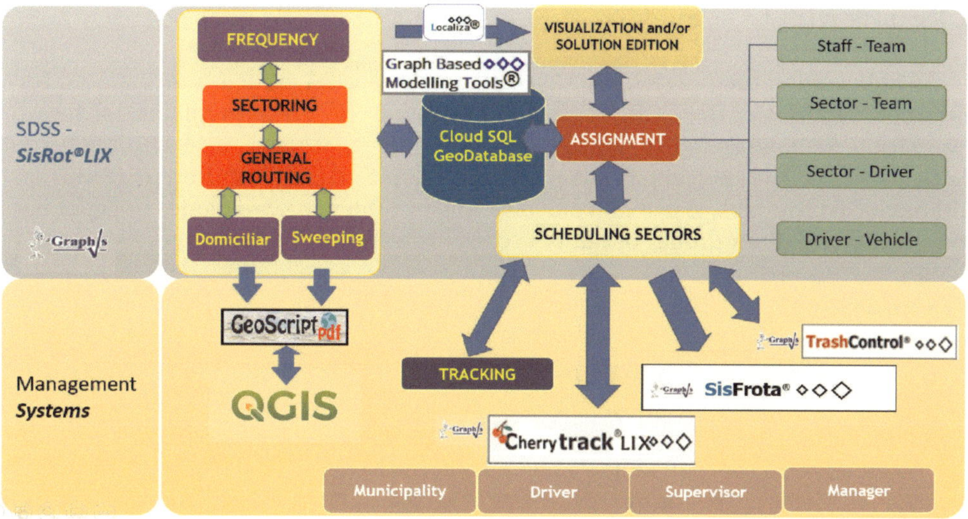

Figure 28.
Spatial decision support system for waste collection logistic planning.

problems in sectoring and routing for heterogeneous fleet, that is informed in the database.

The software is composed of frequency data management and assignment to zones of collection for any type of service, and it contains actual state-of-the-art solvers for solving routes related to general mixed rural postman problem or even general windy rural postman problem, [56, 72].

SisRot LIX® can be used for the operational level. It contains a database to store vehicles characteristics, supervisors, drivers, garis, and teams of collection to schedule them in the related service to be performed daily.

Special reports can be generated by SisRot LIX® with the tool GeoScript®. This tool uses KML files created by the main software that defines zones, sectors, circuits, and routes of the circuits, and draws them in a GIS-based GPS, using different back map resources (Google MAPs, OSM, Bing, etc). All the maps are presented in PDF formats. The KML files generated can be used also by Google Maps or QGIS, as the XLS files that describe the routes street by street.

Once the sectors and routes are defined, they can be sent to another VPS environment that track mobile units running CherryTrack LIX®. This software is composed of four Android mobile applications: Driver, Manager, Supervisor, and Municipality. In the driver unit it is possible to receive the routes (trips) of a sector projected by the SisRot LIX®, and while performing them the software teaches the path as it was projected, facilitating the identification of the work to be done by any driver. Major information about the load per trip, geographical movements, and reported occurrences while running the execution of collection are kept by the software. These information saved in the manager unity that follows each driver in the field, while they are running sector by sector, circuit by circuit. The supervisor app shows the vehicles related only to the supervisor they are allocated. The municipality unity app shows the movements of the vehicle that will attend the citizen home, giving estimates about when it will pass in front to his/her door.

Other tools such as TrashControl® and SisFrota® can be integrated to the VPS to control the loading production and vehicles maintenance standard controls respectively.

6. Results with CCP models

As indicated in the previous sections, all combinatorial models here exposed for different instances of the CCP problem are related to NP-Hard problems. In these approaches the models hold a polynomial number of variables and constraints because there is no need to represent all the partitions related to the clustering process. Although this happens, it is not avoided to solve problem that explodes in exponential complexity with the number of items in the clustering formation.

6.1 IT teams

Here we propose models that are possible to be solved to optimality by state-of-the-art solvers such as Cplex, Xpress, and Gorubi in the case of the IT-Teams Layout, as can be seen in [11].

We presented a solution (**Figure 5b**) for a software factory layout with 121 workplaces and 17 heterogeneous groups, it was obtained in less than 2 seconds using a 10th generation I7-Intel and the Gurobi solver, while this same layout is normally prepared by two persons during 2–3 working days. This impacts directly in better development of teams because people working together do not need to disturb much the other groups in work, meanwhile the interaction of the ordinary groups is facilitated at maximum.

The model proposed does not consider that specific groups has to be close, or even facilitate the dynamic recalculation of groups when the projects are ended and teams are dissolved. These problems remain open, but they can be solved heuristically by the same model if appropriate considerations are made.

6.2 Staff collection and dispatch

Our experience with staff collection/dispatch in the field was with a staff transporter in the state of São Paulo. The transporter works for an industrial food company that had a very confusing schema to collect/dispatch its staff in the state, between 10 municipalities: São Paulo, Guarulhos, Osasco, Carapicuiba, Jandira, Barueri, Itapevi, Taboão da Serra, São Bernardo, **Figure 29a**. The transporter had eight garages where busses departs/returns and the destiny/return at the factory, **Figure 29b**. They had 98

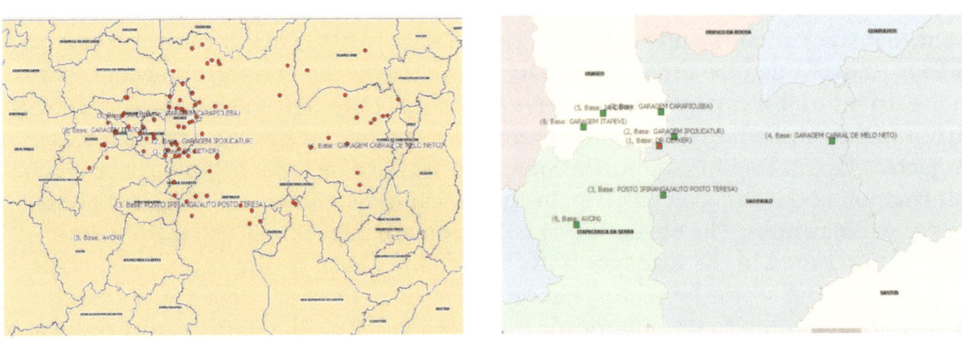

(a) (b)

Figure 29.
Territorial context of the problem. (a) Staff's territorial dispersion. (b) Territorial distribution of contractor's bases and the factory.

workers, which were scheduled along nine entry times and 10 departure times, **Figure 30a**. The transporter offered a fleet of busses with 15, 27, and 48 seats and it had to perform the travels from the first collection to destiny or from the factory to the last dispatch in at most 2 h considering the traffic jams, and the workers must be collected or dispatched at most 800 m far from his home. The workers cannot arrive at the factory late or spend more than 15 min waiting to arrive/departure of any bus.

The indicated problem was proposed to generate the minimum fleet with set of routes that can cover all the shifts with minimum cost of distance and time spent on the activity, while also maximizing the number of bus-stops along the main avenues and streets.

The case was solved by using the methodology here proposed to locate bus-stops as capacitated medians and after applying the routing process over the selected bus-stops. Unhappily our results were not comparable with the transporter because it will not consider the constraint of time for the worker departure. In **Figure 30b** the number of busses in operation per hour of the day is illustrated as calculated by our solution.

Taking a shift as an example, **Figure 31**, we have to perform routes for this shift to pickup and dispatch the staff belonging to it. **Figure 32** illustrates these two steps (**Figure 32a**) of collection and (**Figure 32b**) dispatch of the same set of workers using routes calculated by SisRot Full and sent to Routes that use Google Maps to propose the best route (with better travel time) considering the traffic conditions from arrival/departure to/from factory for a given calculated sequence of visits.

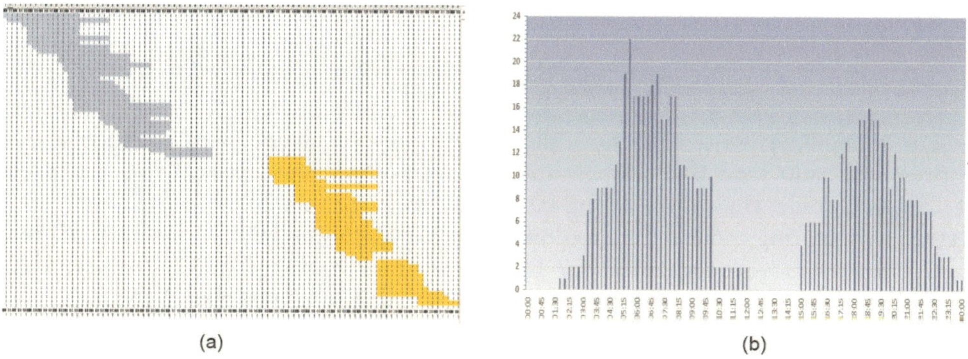

Figure 30.
Staff daily in/out turns and planned fleet to attend them. (a) Staff's labour time distribution. (b) Distribution of the number of buses along daytime.

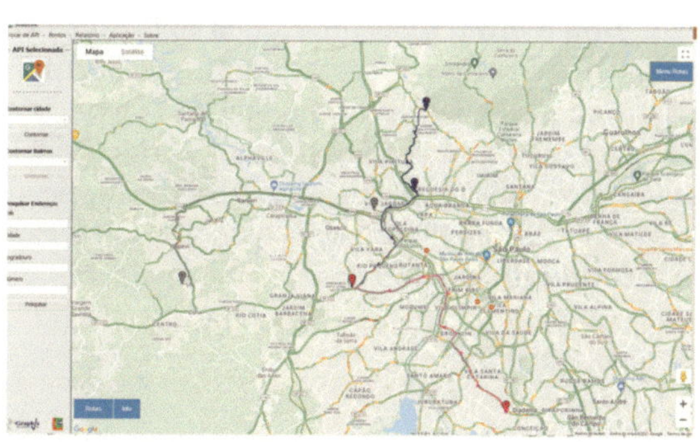

Figure 31.
One shift routes of the staff.

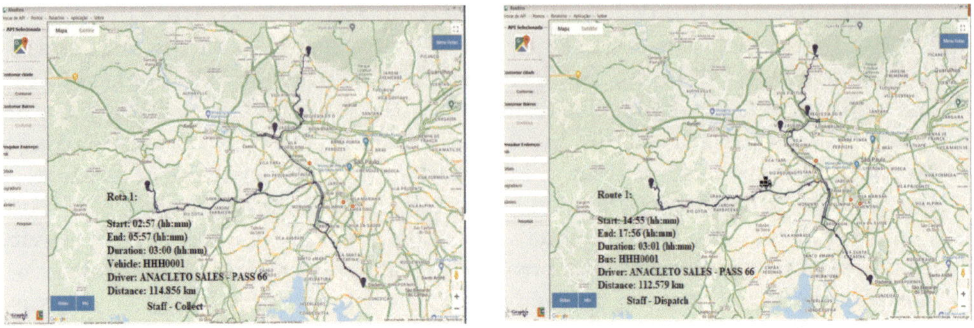

(a) (b)

Figure 32.
Routes of collection and dispatching workers. (a) Route of collecting. (b) Route of dispatching.

6.3 Waste collection

Our experience with CCP models in waste collection is more expressive, because we investigated and developed many cases in Brazil related to the theme using basically SisRot LIX. We divide our main results in domiciliary waste collection and street sweeping.

6.3.1 Results in domiciliary waste collection

For domiciliary waste collection, the sectoring approach returned unexpected and impressive results for CCCP models and solving methodology, even for us. In the city of Campo Grande, the capital of the state of Mato Grosso do Sul (MS) in Brazil, we tested our sectoring models for a region (downtown) covered by five vehicles of 15ton each. The sectors were redefined into four sectors and the routes in total were reduced the distance costs by 28% with the same total time spent for the service, [75].

Most recently, our methodology was tested for the city of Petrópolis in the state of Rio de Janeiro, with a completely different traffic network (many hills and curves) but with the same context. The contractor worked with five vehicles of 8ton in five sectors daily, and after redefining the sectors we achieved for the light days (Tue-Sat) four sectors with 26% of savings in travel distance, and for the heavier days we reduced in 8% the total travel distance performed by the fleet.

In both cases the solution editing tools developed in SisRot LIX were very important and decisive, because the process of sector-routing is performed separately. The manual intervention after the calculation of the sectors is a decision tool related to situations that could be included and are not present at the model's constraints. Problems such as servicing certain places prior to others, duplicate services in different trips on the same sector, side of the service, etc., are some examples of these situations that had to incorporate local decisions and manual intervention.

6.3.2 Results in human street sweeping

The use of CCP models in street sweeping was conducted in an application in the city of São Paulo, for the project PMI-SP organized by the Municipal Secretary of Partnerships during 2018 [69]. The projects were conducted to attend non-scalable services as human and mechanized sweeping, collection in fairs, old furniture and rubble removal, etc.

The CPP models used were conducted by math-heuristics developed by Negreiros et al. [56] and Batista et al. [59], to design sectors basically for human street sweeping. There were 32 sub-municipalities composed by populations between 150 and 400 thousand inhabitants, and diverse coverage areas of sweeping: hills, roads, avenues, and streets on suburbs or downtown, squares, etc. The teams of human sweepers, as reported in subsection 5.3, were composed by two sweepers with a "lutocar", and the sectors to be covered by each team were designed accordingly the frequency of coverage of each one stipulated by AMLURB. Our results revealed that we could develop the same activity in comparison to the previously reported executed plans with 27.7% of reductions in the number of teams and 46.38% in distances performed by the workers.

A "Map of Frequency" of the sub-municipality of Santana-Tucuruvi/SP is presented at **Figure 33**, where among the most evident colors, there are: Light-Green: Daily – Morning; Navy Blue: Wednesday/Saturday – Morning, Green: Monday/ Thursday – Morning, Red: Tuesday/Friday corresponding to the days and regions where the team of sweepers must work. Following, **Figure 34** consists of two figures considering sectoring human street sweepers, in **Figure 34a** it is illustrated in more detail the street map of sweeping frequency to be performed on Tuesdays and Fridays, and in **Figure 34b** it is seen the 21 sectors defined by the HCCCP model to be covered by 21 teams of street sweepers.

As can be seen in **Table 1**, it presents the results of applying Open Sector Arc Routing (OSAR) in the city of São Paulo, in comparison with the solution used by the municipality. **Table 1** represents as Sub-Mun the sub-municipality of the city of São Paulo: BT – Butantã, CV – Casa Verde, PR – Parelheiros, ST – Santana-Tucuruvi, VG – Vila Maria-Vila Guilherme; the term Sectors means the number of sectors defined; Sweep perimeter (km) is the total perimeter that has to be swept by the sweepers; Manual sweep is the number of km per day that the teams perform for the specific solution; Bidding is what it is projected by the municipality; SLix are the values used or projected by SisRot Lix; and finally Gap% means the difference between considerations or solutions.

Table 1 presents the expressive reduction on human resources and distances of work between projects. The cause of this was the effect of the HCCCP-related mathematical methods applied in the practical level, [75].

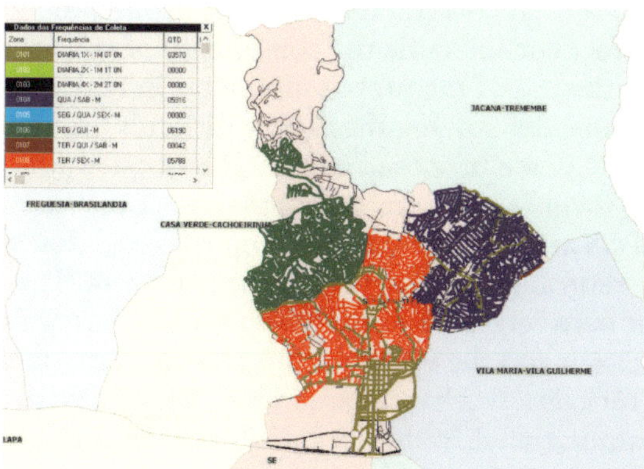

Figure 33.
Map of sweeping frequency for the teams of sweepers.

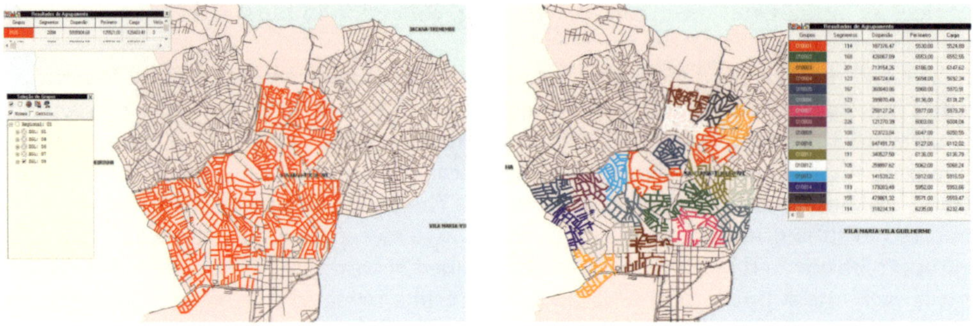

(a) (b)

Figure 34.
Human street sweeper sectoring solution for Santana-Tucuruvi (ST) municipality. (a) Street area for sweeping sectors on Tue/Fri - Morning at ST (b) 21 sweeping sectors for Tue/Fri - Morning at ST.

Sub-Mun	Sectors			Sweep perimeter (km)			Manual sweep - km/day		
	Bidding	SLix	Gap%	Bidding	SLix	Gap%	Bidding	SLix	Gap%
BT	131	121	7.63	763.14	760.29	0.37	275.13	197.72	28.14
CV	82	60	26.83	406.52	407.1	−0.14	199.392	156.57	21.48
PR	36	29	19.44	165.62	166.82	−0.72	61.37	39.62	35.44
ST	152	75	50.66	493.95	321.1	34.99	242.765	148.11	38.99
VG	68	54	20.59	395.24	364.24	7.84	395.744	87.75	77.83
Totals	469	339	27.72	2224.47	2019.55	9.21	1174.40	629.77	46.38

Table 1.
Recorded gains and losses between projects from SP-municipality and GRAPHVS with SisRot LIX.

7. Conclusions

This chapter presents new models for versions of the Capacitated Clustering Problem to IT-Teams layout, logistic distribution, environmental and health applications. Each application needs special constraints that come from basic knapsack and scheduling problems that affect considerable the complexity of the resolution of their instances. We showed the use of the proposed models in real applications and its formal importance to consider solving these very difficult problems.

For MSSC and *g*PCCCP, we proposed convex linear-quadratic alternative models to solve the problems, turning them tractable at least for medium-scaled instances.

We also reported DSS architectures already implemented that are successful in their applications in Brazil, returning high savings to the end users. The architectures presented much differ between real-life applications considering the information needed to support decision and the effort for monitoring plans as they were conceived.

This chapter also reveals a work of a research & development (R&D) team that develops all these computational framework and decision tools, taking into consideration the state-of-the-art in GIS, web services, map services, database services, graph-based modeling, specialized optimization algorithms and solvers for clustering, location, general routing.

As it is shown in the results section, the CCP can be used with wide success for the mentioned applications and could be further used for many different others from the ones here revealed and investigated.

Acknowledgements

We thank Intech Open for the invitation to write this manuscript, specially we also thank the Editor Fausto Pedro García Márquez and Iva Ribic for the kind support during the evolution of this work.

We would like to thank the team of programmers and collaborators from GRAPHVS Ltda: Antônio Honorato Guedes, Bruno Chaves, Cícero Elias Guedes, João Amilcar, and Luiz Prudencio.

We also thank Carlos Higashi, Donizete Calil, Eduardo Reis, André Barufi, and Heraldo Neto for their collaboration in the access to staff collection and dispatching, emergency health systems, and waste collection, respectively.

Finally, our thanks to the researchers and readers of this manuscript, who kindly share with our discoveries, ideas, and great effort to turn this work accessible and worth to the humanity.

Conflict of interest

The authors declare no conflict of interest.

Nomenclature and abbreviations

Bing	BING
CherryTrack	CherryTrack
CherryTrack	CharryTrack Lix
ESRI	ESRI™
Google	Google Maps
Here	Here Maps
GRAPHVS	GRAPHVS™
QGIS	QGIS™
Localiza	Localiza
OSM	Open Street Map (org)
Routes	Routes
Sisfrota	SisFROTA
SisRot	SisRot Full
SisRot	SisRot LIX
SisRot	SisRot SELECTIVE
Trash Control	Trash Control
Webdengue	Webdengue
AMLURB	Autarquia Municipal de Limpeza Urbana de São Paulo
ASRP	Arc sector-routing problem
CCP	Capacitated clustering problem
CCCP	Capacitated centerd clustering problem
COVID-19	Disease provoked by SARs-COVID-19 Virus
CpMP	Capacitated p median problem

*g*CpMP	Generalized capacitated p median problem
*g*CCCP	Generalized capacitated centerd clustering problem
pCCCP	Capacitated p-centerd clustering problem
gMCCP	Generalized multi-capacitated clustering problem
HCSRP	Heterogeneous capacitated sector-routing problem
DATAPREV	Empresa de tecnologia e informaçõμes da Previdência (Brazil)
DSS	Decision support system
DgMCCP	Dynamic generalized multi-capacitated clustering problem
DgMCCCP	Dynamic generalized multi-capacitated centerd clustering problem
EMS	Emergency medical system
ERP	Enterprise resource planning
GBMS	Graph based modeling system
GRP	General routing problem
HCCCP	Heterogeneous capacitated centerd clustering problem
MCCCP	Multi-capacitated centerd clustering problem
MRPP	Mixed rural postman problem
MSSCP	Minimum sum of squares clustering problem
OMRPP	Open mixed rural postman problem
SAMU	Serviço de Atendimento Móvel de Urgência
SERPRO	Serviço federal de processamento de dados (Brazil)
SWM	Solid waste management
WHO	World Health Organization

Author details

Marcos J. Negreiros[1,2*], Nelson Maculan[4], Augusto W.C. Palhano[2],
Albert E.F. Muritiba[3] and Pablo L.F. Batista[2]

1 Universidade Estadual do Ceará, Fortaleza, Brazil

2 Graphvs Ltda, Fortaleza, Brazil

3 Universidade Federal do Ceará, Fortaleza, Brazil

4 Universidade Federal do Rio de Janeiro, Rio de Janeiro, Brazil

*Address all correspondence to: marcos.negreiros@uece.br

References

[1] MacQueen J. Some methods for classification and analysis of multivariate observations. In: Fifth Berkeley Symposium on Mathematics. Statistics and Probability. University of California Press; 1967. pp. 281-297

[2] Hartigan JA. Clustering Algorithms. John Wiley and Sons; 1975

[3] Hansen P, Jaumard B. Cluster analysis and mathematical programming. Mathematical Programming. 1997;**79**(1): 191-215. DOI: 10.1007/BF02614317

[4] Jain AK, Dubes RC. Algorithms for Clustering Data. Prentice Hall; 1988

[5] Jain AK. Data clustering: 50 years beyond K-means. Pattern Recognition Letters. 2010;**31**:651-666. DOI: 10.1016/j.patrec.2009.09.011

[6] Xavier A. The hyperbolic smoothing clustering method. Pattern Recognition. 2010;**43**(3):731-737. DOI: 10.1016/j.patcog.2009.06.018

[7] Xavier A, Xavier V. Solving the min-sum-of-squares clustering problem by hyperbolic smoothing and partition into boundary and gravitational regions. Pattern Recognition. 2011;**44**:70-77. DOI: 10.1016/j.patcog.2010.07.004

[8] Gambella C, Gaddar B, Naoum-Sawaya J. Optimization problems for machine learning: A survey. EJOR. 2021; **290**(3):807-828. DOI: 10.1016/j.ejor.2020.08.045

[9] Pinto R, Ouzia H, Maculan N. New Mixed Integer Non-linear Programming (MINLP) Models for the Euclidean Steiner Tree Problem in Rn. UFRJ-COPPE/ES-Technical Report: ES-769/20; 2020

[10] Linhares MBA. DOIS ENFOQUES PARA A RESOLUÇÃO DO PROBLEMA DE AGRUPAMENTO. (in Portuguese) [MSc Dissertation], UFRJ-COPPE-ES; 2021

[11] Negreiros M, Maculan N, Batista P, Palhano A, Rodrigues J. Capacitated clustering problems applied to the layout of IT-teams in software factories. Annals of Operations Research. 2020. DOI: 10.1007/s10479-020-0378

[12] Mulvey J, Beck M. Solving capacitated clustering problems. European Journal of Operational Research. 1984;**18**:339-348

[13] Negreiros M, Palhano AWC. The capacitated centred clustering problem. Computers & Operations Research. 2006;**33**(6):1639-1663. DOI: 10.1016/j.cor.2004.11.011

[14] Negreiros M, Palhano A, Silva ID. An evaluation of two supply chain models for the newspapers subscribers and bulk milk collection. Revista Produção. 2005;**5**

[15] Liao K, Guo D. A clustering-based approach to the capacitated facility location problem. Transactions in GIS. 2008;**12**:323-339. DOI: 10.1111/j.1467-9671.2008.01105.x

[16] Negreiros M, Xavier AF, Xavier AE, Maculan N, Michelon P, Lima JW and Andrade O. Optimization Models, Statistical and DSS Tools for Prevention and Combat of Dengue Disease. Vol. Chapter 7. London, UK: INTECH; 2011. pp. 115–160. DOI: 10.5772/16857

[17] Boudahri F, Bennekrouf M, Belkaid F, Sari Z. Application of a capacitated centered clustering problem for design of Agri-food

supply chain network. IJCSI - International Journal of Computer Science Issues. 2012;**9**(1): 300-304. DOI: 10.1016/j.ejor.2005.06.074

[18] Chou C-A, Chaovalitwongse WA, Berger-Wolf TY, Dasgupta B, Ashley MV. Capacitated clustering problem in computational biology: Combinatorial and statistical approach for sibling reconstruction. Journal of Advances in Computer Research. 2012; **39**(3):609-619. DOI: 10.1016/j.cor.2011.04.017

[19] Bard J, Jarrah A. Integrating commertial and residential pickup and delivery networks: A case study. Omega. 2013;**41**(4):706-720. DOI: 10.1016/j.omega.2012.09.001

[20] Javadi M, Shahrabi J. New spatial clustering-based models for optimal urban facility location considering geographical obstacles. Journal of Industrial Engineering International. 2014:10-54. DOI: 10.1007/s40092-014-0054-x

[21] Pillai A, Chick J, Johanning L, Khorasanchi M, de Laleu V. Off shore wind farm electrical cable layout optimization. Engineering Optimization. 2015;**47**(12):1689-1708. DOI: 10.1080/0305215X.2014.992892

[22] Moreno S, Pereira J, Yushimito W. A hybrid k-means and integer programming method for commercial territory design: A case study in meat distribution. Annals of Operations Research. 2017;**286**:87-117. DOI: 10.1007/s10479-017-2742-6

[23] Barreto S, Ferreira C, Paixão J, Santos BS. Using clustering analysis in a capacitated location-routing problem. European Journal of Operational Research. 2007;**179**(3):968-977

[24] Buhrmann JH. "The Effects of Clustering on the Medium and Large-Scale Capacitated Location-Routing Problem" [PhD Thesis]. Johannesburg: University of the Witwatersrand; 2016

[25] Deng Y, Bard J. A reactive GRASP with path relinking for capacitated clustering. Journal of Heuristics. 2011;**17**: 119-152. DOI: 10.1007/s10732-010-9129-z

[26] Araújo E, Poldi K, Chaves A. Clustering search applied to the periodic vehicle routing problem: A case study in waste collection. Annals of SBPO - Natal/RN. 2013:1916-1926

[27] Madsen OB. A survey of methods for solving combined location-routing problems. In: Jaiswal NK, editor. Scientific Management of Transport Systems. Amsterdam: North-Holland; 1981

[28] Laporte G, Nobert Y, Arpin D. An exact algorithm for solving a capacitated location-routing problem. Annals of Operations Research. 1986;**6**(9):293-310

[29] Laporte G. Location-routing problems. In: Golden BL, Assad AA, editors. Vehicle Routing: Methods and Studies. Amsterdam: North-Holland; 1988

[30] Nagy G, Salhi S. Location-routing: Issues, models and methods. European Journal of Operational Research. 2007; **177**(2):649-672

[31] Borges Lopes R, Ferreira C, Santos BS, Barreto S. A taxonomical analysis, current methods and objectives on location-routing problems. ITOR. 2013;**20**(6):795-822. DOI: 10.1111/itor.12032

[32] Laporte G. A survey of algorithms for location-routing problems.

Investigación Operativa. 1989;**1**(2): 93-123

[33] Silva Neto IC, Bulhoes T, Subramanian A. Um algoritmo heurstico para o problema de roteamento de ônibus escolares. Annals of LII SBPO, Joao Pessoa/PB. 2020:12

[34] Lopes RB, Barreto S, Ferreira C, Santos BS. A decision-support tool for a capacitated location-routing problem. Decision Support Systems. 2008;**46**(1): 366-375

[35] Muritiba A, Negreiros M, Souza M, Oriá H. Tabu-search with path relinking to the capacitated Centred clustering problem. Expert Systems with Applications. 2022;**198**. DOI: 10.1016/j.eswa.2022.116766

[36] Negreiros M, Maculan N, Xavier AE. (in Portuguese)Projeto Webdengue; Evolução e recursos implementados. UFRJ-COPPE/PESC Technical Report: ES-763/20. 2020

[37] Calixto AB. Dynagraph: Um modelo de edição e representação de graphvs dinâmicos. MPCOMP-UECE-CEFET; 2013 (in Portuguese) [MSc Dissertation]

[38] Chaves B. Uma metodologia de agrupamento e previsão espaço-temporal: Aplicação na Expansão da Dengue e Chikungunya em Fortaleza/CE. MPCOMP-UECE-CEFET; 2019 (in Portuguese) [MSc Dissertation]

[39] Costa DM. (in Portuguese)Uma metodologia iterativa para determinação de zonas de atendimento de serviços emergenciais." [Thesis]. Florianópolis: Eng Produção, Universidade Federal de Santa Catarina; 2004

[40] Galvão R, Morabito R. Emergency service systems: The use of the hypercube queueing model in the solution of probabilistic location problems. International Transactions in Operational Research. 2008;**15**: 525-549. DOI: 10.1111/j.1475-3995.2008.00654.x

[41] Souza RM, Morabito R, Chiyoshi FY, Iannoni AP. (in Portuguese)Análise da configuração de SAMU utilizando múltiplas alternativas de localização de ambulâncias. Gestão & Produção, São Carlos. 2013;**20**(2):287-302

[42] Pons PT, Markovchick VJ. Eigth minutes or less: Does the ambulance response time guideline impact trauma patient outcome? The Journal of Emergency Medicine. 2002;**23**:43-48

[43] Oliveira MJ, Garcia LC. (in Portuguese)A teoria da simulação a eventos discretos na análise do dimensionamento do serviço de atendimento móvel de urgência (SAMU-192). SPOLM. 2005. https://www.marinha.mil.br/spolm/anais/2005

[44] Silva PMS. Análise do serviço de atendimento móvel de urgência (SAMU) de Belo Horizonte via simulação e Otimização.", (in Portuguese) [MSc Dissertation]. Eng Produção, UFMG. 2010

[45] Motta DPS, Lopes RSM, Rodrigues LF. (in Portuguese)Análise de desempenho do SAMU de Ouro Preto e Mariana através de modelos de simulação. Revista de Gestão em Sistemas de Saúde, São Paulo. 2021; **10**(3):296-319. DOI: 10.5585/rgss.v10i3.18710

[46] Larson RC. Hypercube queuing model for facility location and redistricting in urban emergency services. Computers and Operations Research. 1974;**1**:67-95. DOI: 10.1016/0305-0548(74)90076-8

[47] Larson RC, Odoni AR. Urban Operations Research. 2nd ed. Belmont: Dynamic Ideas; 2007

[48] Nogueira Junior LC, Pinto LR, Silva PMS. Reducing emergency medical service response time via the reallocation of ambulance bases. Health Care Management Science. 2014;**19**(1):31-42

[49] Laporte G, Dejax PJ. Dynamic location-routeing problems. Journal of the Operational Research Society. 1989; **40**(5):471-482

[50] Galvão RD, Santibañez-Gonzalez ER. A Lagrangean heuristic for the pk-median dynamic location problem. European Journal of Operational Research. 1992;**58**:250-262

[51] Averbakh I, Berman O, Simchi-Levi D. Probabilistic a priori routing-location problems. Naval Research Logistics. 1994;**41**(7):973-989

[52] Current J, Ratick S, ReVelle C. Dynamic facility location when the total number of facilities is uncertain: A decision analysis approach. European Journal of Operational Research. 1997;**110**:597-609

[53] Rajagopalan HK, Saydam C, Xiao JA. Multiperiod set covering location model for dynamic redeployment of ambulances. Computers and Operations Research. 2008;**35**:814-826

[54] Mourão MC, Nunes AC, Prins C. Heuristic methods for the sectoring arc routing problem. European Journal of Operational Research. 2009;**196**(7): 856-868

[55] Cortinhal MJ, Mourão MC, Nunes AC. Local search heuristics for sectoring routing in a household waste collection. European Journal of Operational Research. 2016:1-12. DOI: 10.1016/j.ejor.2016.04.013

[56] Negreiros, M, Palhano A, Batista AP, and Maculan N. "The heterogeneous sector routing problem". Submitted to EJOR. 2022. p. 54

[57] Ghiani G, Mourão C, Pinto L, Vigo D. Routing in waste collection applications. In: Corberán A, Laporte G, editors. Arc Routing Methods and Applications. Chapter 15 ed. MOS-SIAM Series on Optimization; 2014

[58] Rodrigues AMM, Soeiro J. Waste collection routing—Limited multiple landfills and heterogeneous Fleet. Networks. 2015;**65**(2):155-165. DOI: 10.1002/net.21597

[59] Batista P, Negreiros M, Muritiba A, Palhano A. New framework of metaheuristics for the capacitated Centred clustering problem. Preceedings of XI MIC, Metaheuristic International Congress. 2015

[60] Garcıa-Ayala G, González-Velarde JL, Rıos-Mercado RZ, Fernandez E. A novel model for arc territory design: Promoting Eulerian districts. International Transactions in Operational Research. 2016;**23**:433-458. DOI: 10.1111/itor.12219

[61] Ulusoy G. The fleet size and mix problem for capacitated arc routing. European Journal of Operational Research. 1985;**22**(3):329-337

[62] Logística Ambiental de São Paulo S.A - Loga. Caracterização dos resíduos domiciliares do município de São Paulo Agrupamento Nordeste - 39° quadrimestre Julho/Agosto/Setembro/ Outubro – 2017, CL-NO-10/17. (in Portuguese)

[63] Ecourbis Ambiental Ltda - Ecourbis. Caracterização dos resíduos domiciliares do município de São Paulo Agrupamento Sudeste 3° Quadrimestre 2017 - 39°

Quadrimestre do contrato de concessão. (in Portuguese)

[64] PMSP. (in Portuguese)PGIRS Plano de Gestão Integrada de Resíduos Sólidos da cidade de São Paulo, Comitê Intersecretarial para a política municipal de resíduos sólidos. Prefeitura Minicipal de São Paulo. 2014

[65] SANETAL. (in Portuguese)Plano Municipal de Gestão Integrada de Resíduos Sólidos de Fortaleza. Estado do Ceará", Relatório IV: ACFOR; 2012

[66] PMRJ. Decreto 37.775 - Plano Municipal de Gestão Integrada de Resíduos Sólidos - PMGIRS da Cidade do Rio de Janeiro. 2013. (in Portuguese)

[67] Recicleiros. Recicleiros Cidades (Project). https://recicleiros.org.br/cidades/ [Accessed: March 14, 2022]

[68] Graphvs. SisRot Lix. http://www.graphvs.com.br/?p=65 [Accessed: March 25, 2022]

[69] Graphvs and Barufi. Modelos e Planos de Otimização para a Integração do Sistema Logístico de RSU da Cidade de São Paulo", PMI/SP-2018. São Paulo Parcerias/PMSP; 2018

[70] Corberán A, Sanchis JM. The general routing problem polyhedron: Facets from RPP and GTSP polyhedral. European Journal of Operational Research. 1998;**108**:538-550

[71] Corberán A, Romero A, Sanchis JM. The mixed general routing problem polyhedron. Mathematical Programming. 2003;**96**:103-137

[72] Negreiros M. GTSP based procedures for the mixed rural postman problem. Submmited to Networks. 2022

[73] Prata B. The Multi Capacitated Clustering Problem, Technical Report. Universidade Federal do Ceará; 2015

[74] Araújo KAG, Guedes JB, Prata BA. Hybrid matheuristics for the multi-capacitated clustering problem. RAIRO Operations Research, forthcomming. 2022

[75] Negreiros MJ, Palhano AWC, Reis E. Optimized planning and management of solid waste – Results of application in Brazilian cities (SisRot LIX). In: Solid Waste Management - Recent Advances (ISBN 978–1–80356-327-5). London, UK, forthcomming: InTech Open Books; 2022

Chapter 9

A Conducive Quality Environment Augmenting Practice in Educational Institution

Ashok G. Verghese, Thalavaipillai Sudalaimuthu and Pon Ramalingam

Abstract

A messy workplace has an unstructured environment delay in executing the work. Employees find it difficult to locate the files/document when needed in the messy workplace. The purpose of this proposed study is to apply the practices of 5S in educational institutions. The 5S comprises five phases: Sort, Set in Order, Shine, Standardize, and Sustain. The 5S implementation creates conducive environment in educational sectors that result in effective usage of time and storage space. The 5S implementation generates a significant impact in an educational institution as it relates to the "Swatch Bharat Mission". Clean India Mission is a country-wide campaign initiated by the Government of India in 2014. The 5S is an active tool that helps the educational sectors to improve the global quality of higher education by satisfying the expectations of the investors. Hindustan Institute of Technology and Science serves as the first institution in receiving the Global Certification for 5S implementation in India from TuV (Technical Inspection Association) Rhineland. The execution of 5S creates a dynamic environment for stakeholders of the Institution.

Keywords: educational sector, 5S implementation, Standardize, quality improvement, Sustain

1. Introduction

5S is the implementation of a system at the workplace of an organization. The origins of this method may be traced back to Japan. Hirano and Osada developed two key frameworks separately for understanding and implementing 5S in a corporate context [1, 2]. Sort, Set in Order, Shine, Standardize, and Sustain are the Japanese terms Seiri, Seiton, Seiko, Seikitshu, and Shitshuke, respectively (**Table 1**). 5S is a continuous improvement strategy that focuses on reducing waste, maintaining order, and maintaining cleanliness. The 5S practice is a strategy to increase the employee's performance and morale by establishing and maintaining a quality environment at a workplace [3].

All engineering sectors, including vehicle makers, practice 5S strategies. This method is a tried-and-true method to enhance Quality, Safety, and Productivity. In many businesses, the goal of establishing a quality management system is to increase the efficacy of

1S	SORT	Separation of required and not required essentials work setup. Retain only the essentials that are necessary
2S	SET IN ORDER	Organizing the items in a certain order
3S	SHINE	Implement cleanliness in the workplace
4S	STANDARDIZE	Cultivate certain procedures so that the first 3S is retained.
5S	SUSTAIN	Teach the 5S practices across various places so that it becomes a daily part of life.

Table 1.
Meaning of 5S.

the activity by utilizing existing resources at the appropriate time and at the right place, thereby resulting in a smart solution to the difficulties [4, 5]. According to the author, students assigned to a given task must undergo training to acquire a working knowledge of 5S in the environment where they operate and perform in the organization. It improves the student's ability to understand the approaches early in their career and to provide innovative suggestions for development in their particular workplaces. The paper also defines the execution of 5S activities in a limited enterprise, in which stakeholders are given the chance to put the technique into practice and achieve superior outcomes in terms of layouts and period. The 5S technique is not implemented in all nations in a similar method [6]. The 5S technique is stressed as a strategy for achieving corporate success in the workplace and at home. According to the author, it is only applied as a system for the workplace in nations such as the United Kingdom and the United States, and it is not mandated in other locations [7]. The Chinese's misunderstandings and limits in implementing the 5S technique in manufacturing firms, as well as the consequences after taking the essential phases in the positioning by people. The various types of laboratories ("mechanical, biological, pharmaceutical, and so on") have also adopted this practice and improved their performance [8–12]. The many terminologies of 5S are described in this document, as well as the methodology used for implementation. Initially, displays and training programs were conducted to create 5S awareness among the participants. Second, to assess comprehension, a multiple-choice test was administered with a minimum score of 75% as an eligible criterion; third, based on the assessment results, another cognizance program was arranged, and the response was gathered before the methodology's deployment. The fourth and fifth phases of this task entail putting the first three stages of 5S into action to accomplish the desired results in the various domains [13].

2. Terminology

The following are the explanations for various terms used.

2.1 Workplace

The areas within the educational institution like a lecture hall, laboratory, staff room, mess hall, stores, student hostels, and open areas both inside and outside the building.

2.2 Stakeholders

Administration, teaching staff, parents, students, and alumni.

3. Methodology

Sort, Set in Order, Shine, Standardize, and Sustain is the methodology used in 5S implementation, as depicted in **Figure 1**. The whole campus of HITS (Hindustan

Figure 1.
5S methodology [14].

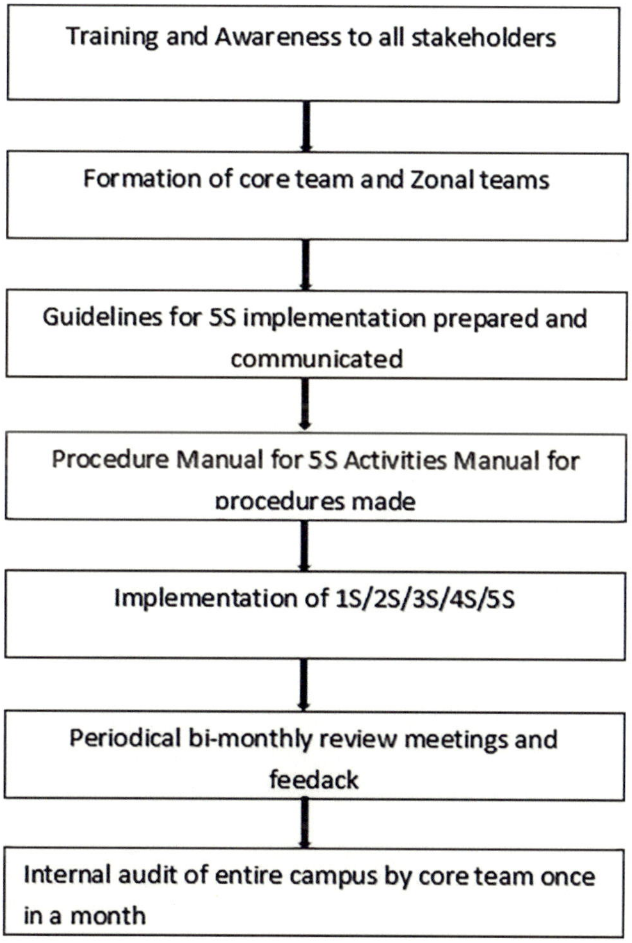

Figure 2.
Flow chart of activities [14].

Institute of Technology and Science) was split into 21 zones. As shown in **Figure 2**, each region was assigned a 5S Supporter and overall branch 5S Coordinators to implement 5S activities step by step.

A training program was held for all personnel in the workplace to raise understanding of 5S ideas. A core team and a steering committee were constituted for 5S implementation. The core team members conducted a workshop on 5S for other stakeholders. **Table 2** contains the specifics of the processes depicted in **Figure 1**.

The success of the training program was determined by assessing the attendees with the help of multiple-choice questions within a given time. In this assessment, all participants need to complete a 1-hour examination in which they were asked to answer 60 questions created at random using software.

Stages	Measures	Guidance
1	All stakeholders will receive training and be made aware of the situation.	• TUV India Pvt., Ltd. provided 5S training to a few selected staff/lab technicians. • 5S core team members conduct an awareness campaign for everyone (staff/students/vendors/sub-contractors). • Posters about 5S were displayed across campus (in open areas, canteens, departments, and dormitories). • All staff and students were administered a 5S objective-type test. • Orientation program for incoming employees and students was conducted.
2	Establishment of a core squad and zonal teams	• The core squad was formed. • 21 zones were identified in the entire campus. • Each zone was allotted champions, department coordinators, and student volunteers. • A 5S organization chart was prepared. • The responsibilities of the facilitators, coordinators, zonal champions, and department coordinators and authority were framed and informed.
3	Guidelines for 5S implementation have been produced and disseminated	• The 5S Coordinator is in charge of supervising the overall implementation and conducting periodic audits to guarantee effective implementation. Deans/HODs were in charge of ensuring that 5S activities were implemented and maintained effectively in their schools/departments. • Everyone in the team was ensured to be aware of the 5S technique. A meeting was conducted with team members to start the execution process. • Periodical meetings were conducted twice a month to examine the progress and report to the coordinator. All documentation and meeting reports were maintained by the coordinator.
4	Implementation of 1S	The cluttered items in the office were sorted into four categories: needed, seldom needed, not needed, and cannot be used. • Required things should be kept close to where they will be used. Rarely used items were stored farther away, and red-tag items were the things not needed or cannot be utilized. The red tag item list is to be prepared and put in the Red Tag Area. • Items with a red sticker must be thrown away with approval from the authority. Record of photos before and after sorting should be maintained. • Frequent reviews should be conducted and necessary corrections should be made.

Stages	Measures	Guidance
5	Implementation of 2S	• Products that are both often and seldom used should be focused on. Objects that are frequently used and items that are infrequently used should be kept in a designated area. • All necessary and infrequently used objects must be appropriately labeled and stored in the designated area. • Any goods that are stacked on the floor should be avoided. • The directions for the numerous workplaces should be properly marked. • Color coding should be used to locate the required item. • A record of before and after photos should be maintained. • An internal audit of the department should be conducted regularly to ensure that things are in their respective places.
6	Implementation of 3S	• Cleanliness should be maintained. • A cleaning schedule should be strictly followed. • A track of the cleaning routine should be checked. • The source of the contamination should be determined and eliminated. • The areas where safety is a concern should be focused on and flaws should be rectified. A random audit should be conducted to ensure compliance.
7	Implementation of 4S	• Sweeping and cleaning everything should be done every day. • Everything around us should be kept in good working order, so we can utilize it whenever we need it. • A cleaning schedule should be established and should be followed. • A record of the cleaning routine should be maintained. • The source of contamination should be determined and eliminated. • On areas where safety is an issue should be concentrated and should be corrected if there are any flaws. A random audit should be conducted to ensure compliance. • If necessary, remedial action should be taken to make changes.
8	Implementation of 5S	• A maintenance system for the first 3S should be created. Procedures, timetables, and practices for the first t3S to be created. • Evaluation of how goods are used and disposed of should be continued. • Checklists should be used regularly to ensure proper housekeeping. • A list of suggestions for implementing the 5S adjustments as a standard operating procedure should be created. • Changes should be reflected in the documentation. • All stakeholders should be aware of the new guidelines.

Table 2.
Stages of 5S with measures and guidance.

The results of this exam were used to determine whether individuals need an additional round of 5S technique awareness training. Then, based on the comments from the participants, another awareness event was conducted. With this readiness, a methodical approach to the execution of 5S activities was taken.

4. Implementation

Each of the tabularized phases was executed at different stages during the implementation process [15]. The audit assessed the feedback at each level and considered the required remedial actions for further enhancement. As a result, continuous progress was necessary to reach the appropriate level for each phase. **Table 3** provides the number of participants for the training program as well as the outcomes of the online exams that were administered. **Tables 4–6** projects the assessment checklists.

Classification	Total No. of participants	No. of participants received scores of 75% or higher	No. of participants scores less than 75%	Percentage passed (%)	Overall pass (%)	Observations
First attempt	312	221	104	70	71	Those who received less than the required score is encouraged to retake the test after gaining more awareness.
Second attempt		105	88	93	96	The outcome of the second attempt
Non-teaching team	121	59	42	53	60	First effort
Second effort		45	40	83	95	Those who received less than the required score is encouraged to retake the test after attending another awareness event.
Other teams						On the third try, the rest of the crew received the required pass %.

Table 3.
Training and outcomes.

SORT: Sort through your belongings to see what's important and what's not. Items that are thought not needed and are not in use should be eliminated from all areas	Improper	Defective	Good	Outstanding	World class
1. Things on the office/staff workplace have been categorized into required and not required categories.	1	2	3	4	5
	Particulars:				
2. Objects in racks have been categorized into useful and non-useful categories.	1	2	3	4	5
	Particulars:				

SORT: Sort through your belongings to see what's important and what's not. Items that are thought not needed and are not in use should be eliminated from all areas	Improper	Defective	Good	Outstanding	World class
3. Objects in cabinets, tables, and file drawers have been categorized into useful and useless categories.	1	2	3	4	5
			Particulars:		
4. Items on the floor have been separated into useful and non-useful categories.	1	2	3	4	5
			Particulars:		
5. To reduce mobility, necessary things have been arranged in the closest area to where they are utilized the most.	1	2	3	4	5
			Particulars:		
6. Have all unnecessary objects been removed from the workplace?	1	2	3	4	5
			Particulars:		
7. Work agreements for the aforementioned are acknowledged. Every employee is aware of where they may locate them.	1	2	3	4	5
			Particulars:		

Improper: Proof is not shown.
Defective: Random evidence is provided.
Good: Implemented and mostly proven.
Outstanding: Proven and entirely implemented.
World class: Continually identifying the methods for improvement.

Table 4.
Audit checklist on 1S.

SET IN ORDER: There should be a place for everything, and everything should be easy to find.	Improper	Defective	Good	Outstanding	World class
1. Needed item locations are labeled, and objects are stored in designated areas.	1	2	3	4	5
			Particulars:		
2. Required amounts for essential objects, including things in desk drawers and bookshelves, are determined.	1	2	3	4	5
			Particulars:		
3. Removable item locations are identified, and things are placed in the proper positions.	1	2	3	4	5
			Particulars:		
4. Indicators and visual controls have been recognized	1	2	3	4	5
			Particulars:		
5. System to assure crucial item reorder levels.	1	2	3	4	5
			Particulars:		
6. The substances of drawers and lockers are clearly labeled (a new person should be able to locate them without assistance).	1	2	3	4	5
			Particulars:		

Table 5.
Audit checklist on 2S.

SHINE: Maintain a clean and ready-to-use work environment. Regularly inspect to verify that the sort and order are maintained.	Improper	Defective	Good	Outstanding	World class
1. Workspaces and equipment are continuously supplied and arranged following 5S agreements and timetables.	1	2	3	4	5
			Particulars:		
2. Every day, members of the workgroup adhere to the 5S agreements.	1	2	3	4	5
			Particulars:		
3. All staff members document the sources and frequency of 5S issues as part of their regular job.	1	2	3	4	5
			Particulars:		
4. Dust and dirt are removed from all surfaces.	1	2	3	4	5
			Particulars:		
5. Checklists are used to identify ongoing Shine responsibilities and to keep track of their progress.	1	2	3	4	5
			Particulars:		

Table 6.
Audit checklist on 3S.

5. Results and discussions

5.1 1S – Sort

The following enumerates the organization of objects at the workspaces:

1. It's not working and cannot be used.

2. Items that are useful but aren't required at work.

3. Items that are required for work but aren't in working order.

4. Items that must be used at work and are in good operating order.

Pictures before and after Sort are taken (**Figure 3**).

The sorting technique was implemented in the workplace. **Figure 3** shows a clear representation of the implementation of 5S. Initially, the required and not required items were not sorted. Later the things as required and working conditions were identified and sorted to access the materials effectively.

5.2 2S: Set in order

The objects in defined areas were organized (**Figure 4**) to make them easy to find and named so that anybody can find them. The goal is to offer a handy, tidy, and secure location for everything and to arrange everything so that way recovery is simple. When organizing, the items regularly used should be maintained at the point of use, while items that are infrequently used should be stored away from the work area.

Figure 3.
Sort (1S in the workplace): (a) before implementation, and (b) after implementation.

Figure 4.
2S – Set in Order.

Addresses are created and labels are applied to places, as shown in **Figure 5**. Color-coding and marking are also shown in **Figure 5**. Setting things in order is crucial since it improves the essential actions in day-to-day operations while also saving time. Before and after implementation, photos are taken. By putting things in the right sequence, retrieval time is reduced, resulting in increased efficiency and effectiveness for both the person and the system as a whole. Before and after implementation, photos are taken.

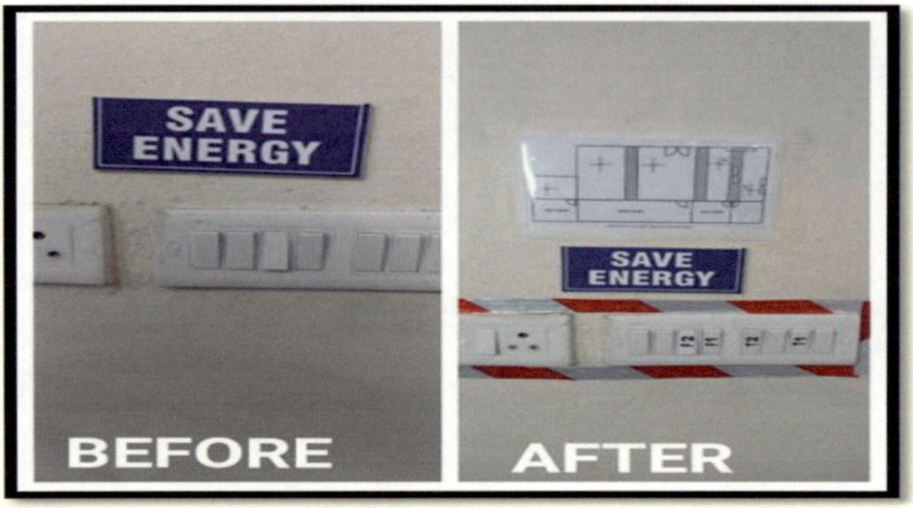

Figure 5.
2S – Labeled switches in the campus.

By putting things in the right sequence, retrieval time is reduced, resulting in increased efficiency and effectiveness for both the person and the system as a whole.

5.3 3S – Shine

The whole workplace is cleaned in this process. Shine's goal is to transform the workplace into a clean, bright atmosphere where everyone enjoys working. Cleaning should become embedded in everyday work routines, ensuring that tools, equipment, and work locations are always available for use. Cleaning and methodically inspecting the system allows discrepancies, issues, and opportunities for improvement to be easily identified [5]. **Figure 6** provides a representation of the laboratory floor with a floor mark to ensure open aisle space.

Figure 6.
Implementation of 3S in institution.

Figure 7 depicts a graphic representation of a potentially dangerous region (backside of door opening).

Figure 8 [16] depicts motivational messages on the staircase to motivate everyone in the office. **Figure 9** [16] depicts an electrically hazardous location. **Figure 9** depicts the correction step to guarantee the safety of the wires that are dangling. A 5S Corner is set up in the work area to raise awareness and serve as a communication board for tracking progress.

5.4 Standardize (4S)

The crew identified approaches to establish better workplace behaviors as a standard and visible management throughout this phase of implementation. Standard operating procedures, checklists, and templates are developed and implemented to maintain a visually instructive, clean, and organized work environment. The

Figure 7.
Safety sign indicator.

Figure 8.
Staircase exhibiting slogans.

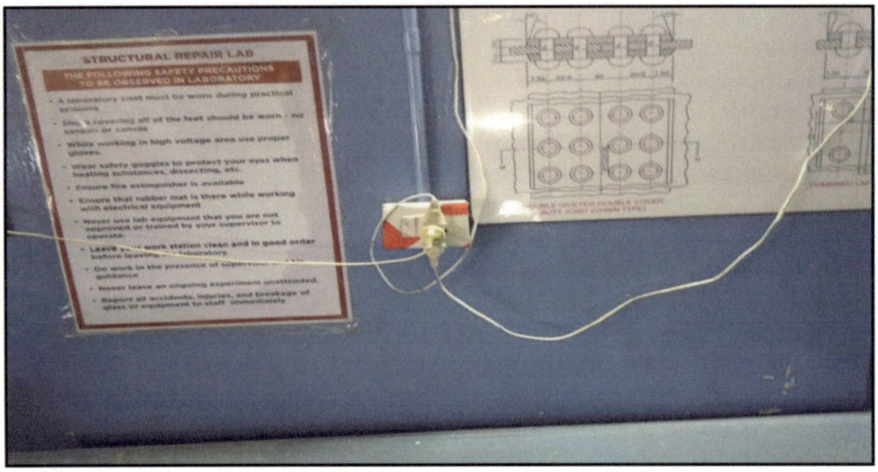

Figure 9.
Safety a concern.

S.no	No. of regions	No. of proposals	No. of proposal nominated	Short tenure	Long tenure	Observations
1	21	80	20	16	4	All short-term goals are accomplished. Long-term goals are in progress.

Table 7.
Details of suggestions from stakeholders.

objective of 4S is to establish the best methods, and each member of the team followed them in the same way. The first 3S including Sort, Set in Order, and Shine are all included within Standardize.

5.5 Sustain (5S)

Sustainment entails self-discipline and the development of a habit of following the preceding 4S. Internal and external audits were performed regularly to ensure that the system's efficacy and continual improvement are being monitored. Periodic walks around campus are taken to ensure that the campus is litter-free and to identify any deficiencies or areas for continuous progress. Efforts to maintain 5S implementation are acknowledged and satisfied. The 5S initiative is an element of the new employee orientation program as well as the new student orientation program (**Table 7**).

The completed short-term goals are as follows: (a) dustbins, (b) window lock in proper condition, (c) direction boards, (d) proper electrical switches, (e) fire extinguishers, (f) proper water stagnation, (g) proper maintenance of washroom, (h) management of files, (i) proper shelf and cupboards, and (j) emergency exit. The ongoing long-term goals are as follows: (a) proper painting, (b) facilitating lab in same blocks, (c) pedestrian walking space, and (d) prevention of leaking floors.

6. Conclusions

A vital stage in 5S implementation is raising awareness and providing training to all stakeholders. All stakeholders, including administrative officials, staff, parents, students, and service agencies, have participated and contributed for efficient and successful system. The use of 5S has resulted in a clean and organized atmosphere. Staff and student self-discipline have significantly improved. The Hindustan Institute of Technology and Science is the first educational institution in India to gain world-wide accreditation for 5S implementation from TuV Rhineland. All management staff, faculty members, and students have been motivated to achieve a clean, safe, and thought-nurturing workplace as a result. Several non-value operations added to the existing procedure will be recognized in the future and appropriate remedial actions will be taken. In addition, time management for various operations will be monitored, which will help to increase overall efficiency.

Abbreviations

TuV	Technical Inspection Association (Technischer Uberwachungsverein)
5S	Sort, Set in Order, Shine, Standardize, and Sustain

Author details

Ashok G. Verghese, Thalavaipillai Sudalaimuthu* and Pon Ramalingam
Hindustan Institute of Technology and Science, Chennai, India

*Address all correspondence to: tsmuthu@hindustanuniv.ac.in

References

[1] Lepeley MT. Human-Centred Management: 5 Pillars of Organizational Quality and Global Sustainability. London: Routledge; 2020

[2] Parkinson S, Lowe C, Keys K. Professional development enhances the occupational therapy work environment. British Journal of Occupational Therapy. 2010;**73**(10):470-476

[3] Ho SKM. Japanese 5-S-where TQM begins. The TQM Magazine. 1999;**11**:311-320. DOI: 10.1108/09544789910282345

[4] Zhang H. How to Carry out "5S". Activities in the student dormitory management. Journal of Human Economic Management. 2005;**2005**:16

[5] Chi HL. 5S Implementation in Wan Cheng Industry Manufacturing Factory in Taiwan. Menomonie, WI, USA: University of Wisconsin-Stout; 2011

[6] Filip FC, Marascu-Klein V. The 5S lean method as a tool of industrial management performances. In IOP conference series. Materials Science and Engineering. 2015;**95**:012127

[7] Lixia C, Bo M. How to Make 5S as a Culture in Chinese Enterprises. China: School of Economics and Management; 2011

[8] Altamirano, R.J., 2013. Aplicación de la metodología japonesa de calidad 5S para optimizar las operaciones en el laboratorio de mecánica de patio de la Universidad de las Fuerzas Armadas – ESPE. Universidad de las Fuerzas Armadas – ESPE, Sangolquí, Ecuador

[9] Ananthanarayanan KRM. Application of 5S management system in NDE laboratory. In: Proc: National Seminar on Non-Destructive Evaluation. Hyderabad, India; 2006

[10] Chitre A. Implementing the 5S methodology for lab Management in the Quality Assurance lab of a flexible packaging converter. In: Master of Science Degree in Technology Management. WI, USA: The Graduate School University of Wisconsin-Stout Menomonie; 2010

[11] Mallick A, Kaur A, Patra M. Implementation of 5S in pharmaceutical laboratory. IJPRBS. 2013;**2**(1):96-103

[12] Pentti O. Applying the Lean 5S Method to Laboratories and Prototype Workshops. Finland: Turku University of Applied Sciences; 2014

[13] Asmit S, Hernadewita H, Ismail A. Implementation of 5S principles in Genset's house palm plantation. Journal of Applied Research on Industrial Engineering. 2018;**5**(2):122-130

[14] Ashok G, Viswanathan K, Ramalingam P. 5S Implementation in Workplace - a Conducive Environment Enhancing Motivation in Educational Institution. In: Proceedings of the International Conference on Industrial Engineering and Operations Management. Paris, France: IEOM Society International; 2018

[15] Kumar K, Kumar S. Steps for implementation of 5S. International Journal of Managment, IT and Engineering. 2012;**2**(6):402-416

[16] Available from: https://issuu.com/hindustanuniversitypublications/docs/hcellence_magazine_2020-2021

Chapter 10

Assessment of Medical Equipment Maintenance Management

Călin Corciovă, Robert Fuior, Doru Andrițoi and Cătălina Luca

Abstract

Today's modern hospital is highly dependent on different types of medical equipment to help diagnose, monitor, and treat patients. Medical equipment maintenance is important to reduce costs, reduce patient dissatisfaction, treat the patient in a timely manner, and reduce mortality and risks during patient care. Good maintenance management is important to have well-planned and implemented programs through which hospitals can minimize medical device failures or other problems with the operation of medical equipment. Medical equipment plays an important role in the hospital system; therefore, the acquisition, maintenance, and replacement of medical equipment are key factors in hospitals for the implementation of the health service. Thus, in order to ensure the quality of medical devices for the provision of medical care, it is imperative to evaluate the safety of using hospital maintenance management. In order to achieve these goals, hospitals must develop checklists that identify the state of performance of medical equipment maintenance. It is essential for clinical managers and engineers not only to increase the capacity of the hospital but also to predict the risks of sudden failure. Given the lack of unique and comprehensive maintenance management checklists, the current goal is to design and develop medical equipment maintenance management checklists.

Keywords: maintenance strategies, healthcare technology management, risk, prediction, medical devices

1. Introduction

Medical equipment is a product that can directly affect human lives. They have been and are considerable investments and in many cases have high maintenance costs. That is why it is important to have a well-planned, managed maintenance program that can keep medical equipment in a reliable, safe, and available medical service. In addition, such a management program extends the life of the equipment and minimizes the cost of its maintenance [1].

A modern medical equipment maintenance strategy includes periodic inspection procedures to which preventive maintenance (PM) and corrective maintenance (CM) are added when necessary. Performance inspections ensure that the equipment operates correctly within the limits set by the manufacturer or according to the standards in force for that type of equipment. Safety inspections ensure that the equipment is safe to use for both patients and operators. Preventive maintenance (PM) aims to extend the life of the equipment and reduce the downtime of medical equipment [2].

In addition, there may be some hidden issues that can be easily detected during a scheduled inspection. The technical inspection procedure of medical equipment only ensures that the device is in good working order at the time of inspection and cannot eliminate the possibility of a malfunction during future use. It must be borne in mind that by their nature, electrical and mechanical components can be damaged at any time. Corrective maintenance (CM) restores the operation of a defective device and allows it to be put back into operation in optimal parameters.

2. Maintenance program planning (inventory and resources)

Planning a maintenance program requires more effort to establish a comprehensive medical equipment management program. There are a number of critical factors to consider:

1. Inventory – types and models of medical equipment to be followed by technical staff and the steps by which they are specifically included in specific maintenance programs;

2. Methodology – identify the method by which maintenance will be provided to the medical equipment included in the program;

3. Resources – the financial, technical, and human resources included in the program.

In the process of planning a maintenance program, it is essential to determine the types of medical equipment they need to be included in a maintenance management program. This fact will depend on the types of medical services to be provided covered by the program, ranging from primary care clinics to hospitals as well as the range of equipment in them. The clinical engineering department of the hospital is the one that has to identify and select the medical equipment to be included in the inventory and to include it in the maintenance program [3].

A maintenance program can be implemented in several ways, the variety of methodologies available at a given time must be taken into account. For example, the healthcare provider may enter into service contracts with device manufacturers, independent service organizations, or a combination of both, in addition to the assistance that their own technical staff can provide.

As for the resources needed for maintenance, they are difficult to predict. This requires a well-maintained maintenance history, personnel requirement calculations, test equipment as well as their technical knowledge to repair defective equipment. Outside suppliers are required for the maintenance of complex equipment. Maintenance often requires access to equipment that can be difficult to obtain due to budget constraints and purchasing difficulties, especially when buying from abroad (**Table 1**). In order to meet such challenges, it is important to consider the financial advance, technical, and human resources required to properly carrying out the planned activities [4].

A maintenance schedule is also based on a defined number of physical resources. This category includes workspace, tools and test equipment, consumables, spare

	Initial cost	**Operating cost**
Technical resources	Space, tools, test equipment, and computer resources	Utilities, operation, maintenance, and calibration
Human resources	Recruiting and initial training	Salaries and continuing education
Maintenance	Not applicable	Service contract, spare parts, and materials

Table 1.
Financial resources required for a maintenance program.

Medical device	**Test equipment required**
All electrical equipment	Electrical safety analyzer
ICU monitors and ECG machines	Simulators/arrhythmia simulators
Defibrillators	Defibrillator analyzer
Electrosurgical units	Frequency electrosurgical analyzer
Ventilators, heart-lung machine, and anesthesia machine	Pneumatic tester and pneumatic flow meter
Noninvasive blood pressure monitors	Noninvasive blood pressure simulator
Anesthesia machines and ventilators,	Gas flow meters

Table 2.
Test equipment for some medical device category.

parts, and maintenance and service manuals required to perform maintenance. Various testing tools and equipment are required to perform PM and/or CM procedures, depending on the type of medical equipment. It is possible to perform many PM and CM with a basic set of electronic service instruments and test equipment (e.g. temperature meter, voltmeter, power indicator, oscilloscope, and electrical safety meter) (**Table 2**). However, all these involve additional costs that must be provided in the financial and technological management.

3. Operational management (developing or changing inspection and preventive maintenance procedures)

Fortunately, in hospitals, we see more and more technology. And this can only make us happy, proving to us that medicine is advancing, so the diagnosis is made earlier and more precisely, the treatment is more personalized and more effective. Overall, technology allows us to better care for the patient for his/her benefit, but also for the doctor, who has powerful tools to practice his/her profession. At the same time, the presence of these increasingly sophisticated medical equipment comes with the need for medical engineers in hospitals who know how to use this equipment, to identify when they are not used correctly or when they no longer work at optimal parameters.

For a patient, the staff of a hospital means doctors, nurses, and sisters, and it is normal to be like that, because only with them they interact but, a hospital in order to operate continuously, 24/7, in safe conditions, there is a whole team of technicians behind it, which makes this possible. Someone has to make sure that the medical equipment, the ventilation systems, the medical gas systems, and the electrical network are working properly, and these are the bioengineers, the clinical engineers, and the medical engineers.

Operational management involves making very diverse decisions, which can be classified into two broad categories: strategic and tactical. Strategic decisions have a longer time horizon and are less structured than tactical ones. They focus on the entire organization, drawing certain lines and directions, while tactical decisions are more focused on departments, teams, and issues. Device management takes place within a generalized business model, so clinical engineers work in collaboration with the financial and procurement departments of the institution to ensure efficient management of technical equipment, but also assistance policies are properly implemented through the efficient use of financial resources [5]. Consequently, the role of clinical engineering includes the development of equipment support strategies, the management and periodic review of these strategies, as well as the management of both fiscal resources and specialized personnel. Various problems may occur with medical devices resulting from device malfunctions, equipment operation problems, or malfunctioning devices [6, 7].

The first maintenance policies developed consisted of interventions on the devices that worked until their accidental shutdown (breakdown) due to wear and tear or due to the occurrence of malfunctions. The intervention is considered satisfactory as long as the device/system is operating at a minimum acceptable level (reactive maintenance) [8].

The development and increase in the complexity of medical equipment have led to the modernization and updating of maintenance techniques and policies. The preventive and predictive activity makes it possible to plan the shutdown, prepare the intervention team, ensure the necessary spare parts, and respectively, reduce to a minimum the parking time for repair [9]. Predictive maintenance is a qualitative leap forward in a modern maintenance system, regardless of the scope or specifics of production, as it provides all the information needed to:

- early detection of defects;
- their location;
- fault diagnosis; and
- calculation of the safe operating time of the machine.

Many healthcare providers have now implemented continuous quality improvement programs based on innovation in the way care is provided in order to improve safety, control costs, and make these services more accessible and effective for patients. Staff involved in operational management also use risk management methodologies to optimize hospital resources to provide a healthcare technology management program that focuses on ensuring that the hospital's clinical work can be performed safely and cost-effectively.

The lack of specialized technical staff, including medical bioengineers and clinical engineers in hospitals, as well as an inefficient maintenance system are the major causes that determine incidents of all kinds. This has been highlighted more than ever, during the pandemic, when hospitals and all their medical equipment were and still are overburdened. During this time, more than ever, we have realized that the rules and standards of hospital care (including the care of medical equipment, medical gas networks, ventilation, electrical networks, etc.) need to be improved, and maintenance policies need to be optimized as quickly as possible.

4. Risk-based biomedical equipment management program

Healthcare is one of the largest industries in the world, with a high degree of diversity in terms of therapeutic activities and how they are performed. There is compelling evidence that while healthcare brings enormous benefits to all people, the frequency of errors and unwanted events is increasing, directly related to the development of innovation in biomedical technologies. Understanding and ensuring the safety of medical care is an extreme challenge in health system management [10].

For departments and organizations that target medical devices, whether they are manufactured or used, one of the key goals should be patient safety and risk management. Risk management is a complex process of identifying, analyzing, and responding to potential risks, through a documented approach, which uses material, financial, and human resources to achieve objectives, aiming to reduce their exposure to losses [11]. Thus, internal control is directly associated with risk management, because, through the measures taken, a functional framework is reasonably ensured that allows that entity to achieve its objectives [12].

The risk appetite of the organization must be clearly articulated in the policy, and this can be informed through legal and financial issues. Policy should also clearly define roles and responsibilities in risk management. There must also be a periodic review process in which each risk is reviewed to ensure that control measures are effective and that the residual risk is properly classified. The policy should set out the process, methods, and tools used to manage the risks within the organization.

Risk management is a cyclical process, which takes place throughout the course of an activity and involves several stages of work as shown in **Figure 1**.

4.1 Risk identification

The first step in the risk management process involves focusing efforts on identifying all possible sources of risk that could affect in any way the development of the project or

Figure 1.
Risk management cycle.

activity analyzed. Effective risk management assumes that risk identification is an ongoing process that allows the entity to connect to the process of change and adaptation. An efficient risk management process at the level of the entity must also take into account the priorities of the institutions under coordination/subordination or under authority, which contribute to the achievement of the objectives of the respective entity [13].

The sources of risk come from both inside and outside an organization that provides healthcare services. External sources are those sources of risk that are the result of events outside the organization under analysis. The main factors that can influence the external risk environment and that need to be taken into account are as follows:

- regulations and/or legislation – each healthcare provider must identify those laws and regulations under which they operate and which define the limits of action of this entity;

- modifying/updating the objectives – in some situations, the treatment of some risks by the managers of the organization is influenced by the external decisions that influence the own activities;

- sometimes, budget cuts can affect the achievement/limitation or stoppage of some professional activities or the number of employees.

Due to the fact that in the case of external risk no preventive measures can be taken, the only way to act is by insurance. Internal sources are the result of events within the organization. These sources of risk can be controlled. In this category, we can distinguish the risks of using technology equipment, the risks of specialized labor, or the risks associated with organizational management. These risks can be prevented by simply eliminating the sources that produce them, which is possible due to the fact that they are generated by the activity of the organization, so they come from within it.

4.2 Risk assessment (impact/probability)

There are two main categories in the risk analysis process:

- qualitative risk analysis;

- quantitative risk analysis.

The results of the qualitative risk analysis are less accurate, as they are more indicative than precise. In the qualitative analysis, the Probability Impact Matrix technique can be used, a technique that combines the two components of risk, thus presenting an overview of it [10]. This method can be applied at several levels, with varying degrees of difficulty. Most healthcare organizations use a 535 risk matrix based on Australian standards AS/4360: 2004, where the two axes correspond to the scales for consequences, sometimes called severity and probability as illustrated in **Figure 2** [14].

If these results are not satisfactory, the risk management also provides the quantitative analysis which presents results in numerical form as a result of the calculations made. If the risk analyst cannot give an accurate probability of such an event occurring, he/she can instead calculate the size of the losses or depreciations generated.

In this sense, risk management has developed a series of calculation methods and techniques such as [15]:

		Consequence				
		Negligible 1	Minor 2	Moderate 3	Major 4	Catastrophic 5
Likelihood	5 Almost certain	Moderate 5	High 10	Extreme 15	Extreme 20	Extreme 25
	4 Likely	Moderate 4	High 8	High 12	Extreme 16	Extreme 20
	3 Possible	Low 3	Moderate 6	High 9	High 12	Extreme 15
	2 Unlikely	Low 2	Moderate 4	Moderate 6	High 8	High 10
	1 Rare	Low 1	Low 2	Low 3	Moderate 4	Moderate 5

Figure 2.
Probability impact matrix examples.

- SWOT analysis (**S***trengths,* **W***eaknesses,* **O***pportunities and* **T***hreats)*: in practice, a few risk analyses have been shown to be effective when using this simple and well-known method.
- Concatenation principle: by concatenating the risks, a chain of risky events is identified, and these will be analyzed together.
- Script technique: this technique involves describing one or more possible ways of conducting an event from a concrete situation.
- Expected value analysis.
- Decision-based tree analysis.
- Monte Carlo simulation method.

4.3 Reaction to risk/attitude toward risk

Measures should be taken to reduce the probability (possibility) of occurrence of the risk and/or to reduce the consequences (impact) on the results (objectives), if the risk is materialized. Risk response is a reduction in risk exposure if it is a threat.

If disposal is not feasible, control measures should be put in place:

- Elimination or avoidance
- Replacement
- Control of risks at source
- Separation and isolation

- Safe working procedures
- Training, instruction, and supervision
- Personal protection
- Other considerations: social assistance facilities, first aid, and emergency procedures

Lack of control can compromise the entire risk management process.

4.4 Risk monitoring, review, and reporting

In general, the risk management framework should not be seen as a single task but should be continually reviewed to ensure that it remains fit for purpose. There are a variety of technical standards for medical devices designed to ensure the consistency of a device's safety throughout its life [16, 17]. These should be seen as a starting point in the risk profile and are the basis of the medical device directives: directive on implantable active medical devices (90/385/EEC), Medical Devices Directive (93/42/EEC), and directive on in vitro diagnostic medical devices (98/79/EEC). Another useful standard is BS ISO 31100 which sets a framework for incorporating risk management into any organization [18].

5. Work order system for corrective maintenance

A very important component in the management of medical technologies is represented by the realization of a maintenance program in which to take into account the characteristics and failures of medical equipment. At the moment of including such a maintenance strategy, a distinction will have to be made between older medical devices and high-tech devices, as they cannot be effectively managed if the same maintenance strategies are used [19]. The World Health Organization has issued guidelines for the application of corrective maintenance at the hospital level. Corrective maintenance is actually a whole process from identifying the defect to repairing the device and putting it back into operation.

5.1 Troubleshooting and repair

The first step in the corrective maintenance process is to report a fault/incident to a user. Also, a malfunction can be found when a medical technician/bioengineer from the Department of Clinical Engineering performs the periodic technical inspection. In order to make the maintenance process more efficient, the first step is the realization of the finding and then the corrective maintenance is initiated. A reduction in the failure time of the defective medical device can be achieved when the medical technician/bioengineer performs some corrective maintenance steps himself/herself and uses internal expertise or external service providers. This corrective maintenance may be accomplished at various levels:

- component level: old generation equipment requires isolation of the fault by troubleshooting and repairing the level of components. In the case of new generation medical equipment, electronic devices, repairing the level of components can be time-consuming and difficult. In addition, this type of maintenance is not feasible, so the repair of the board level or even the system level is applied.

- board level: in the case of modern electronic equipment, the faults of a certain circuit board are isolated, and the entire board is replaced.

- device or system level: in the case of modern equipment, troubleshooting and repairing at the plate level are sometimes difficult and time-consuming. Therefore, if the cost of a repair exceeds 60% of the value of the purchase of new equipment, it is more profitable to replace the entire device or subsystem.

The type of corrective maintenance applied at the hospital level is dependent on a number of cumulative factors including the availability of financial, physical, and human resources, as well as the urgency of a particular request for repair. One strategy that can be considered is the application of repairs at the device level, in emergency situations and when more time is available, the repair can be used at the plate level or at the component level. If it is proposed to repair the component level, parts may need to be replaced.

Replacement can be done with specialized parts from the manufacturer or with spare parts recovered from malfunctioning or obsolete equipment (only after the announced risk assessment).

5.2 Factors affecting equipment failures

In the event of an unforeseen fault, environmental factors must also be considered, such as stabilizing the power supply sources by using voltage regulators, installing uninterruptible power supplies (UPS), using surge protection devices and avoiding connection. Another important environmental aspect is the interaction of medical devices with other utility systems (e.g. medical gas and vacuum systems, temperature and ventilation control systems, water supply, and information technology).

There must be a permanent collaboration with the other categories of engineers in the organization in order to optimize the capacity of the utility systems so that the medical equipment works in optimal parameters. The environment in which medical equipment is used should be controlled in terms of temperature and humidity. There are situations in which some medical equipment is designed to be used in accordance with the specific climate of a country/region. In this case, the maintenance procedures in a particular country or region are adjusted according to these local factors. Other important environmental factors are the age and condition of medical equipment or old facilities built to old standards that are not applicable.

5.3 Inspection and return to service

The actions required after the repair is completed include recalibration processes and a performance and safety inspection. These activities are essential for measuring device performance. Once these activities are completed, the medical equipment will be returned for use to the patient care.

5.4 Reporting

In the case of corrective maintenance, a decisive role is played by the periodic technical verification, part of the preventive maintenance process. The technical file of the medical device in which any intervention made on the medical device is recorded is a reference document in corrective maintenance that helps to make decisions. Existing data in the worksheet includes the actions performed, the parts replaced, and their cost helps to identify if or when the parts need to be replaced again and helps to explain the condition of the parts during the current inspection.

5.5 Safety

An important aspect in the corrective maintenance process is the consideration of safety aspects. In this case, procedures must be included regarding the safety of the technical staff during maintenance, the safety of the user after maintenance, and the general control of infections. In order to increase the safety of service personnel, it is necessary to train and use personal protective equipment and knowledge of techniques that will allow technical personnel to work safely in dangerous conditions.

Following maintenance, in particular following procedures which could have affected the safety features of a medical device, the technical staff must check that the device is safe to use, mechanically and electrically. Particular attention shall be paid to the electrical safety of medical devices so that they are earth-proof and leakage current is measured to ensure that they are within the applicable limits. (In the absence of electrical safety test equipment, technical personnel must rely on careful repair techniques and simple electrical tests to verify the integrity of the device). Physicians should be advised to check the settings of the device and perform basic operational checks before using the device with patients [20].

An important conclusion we can draw is that the best maintenance strategy is to apply the mixed maintenance strategy. This process is due to the fact that we are talking about a very large number of medical equipment and complex technologies. The maintenance process must be divided between the parts: internal maintenance is adopted for surgical lamps, sciatic lamps, and telemetry devices. In the case of critical devices, maintenance is covered by full-risk agreements with authorized manufacturers or service centers.

External maintenance is adopted for anesthesia machines, mechanical ventilators, electrocardiographs, patient monitors, and surgical tables [21].

6. Managing a maintenance program in a hospital

Good management of medical technologies at the hospital level can minimize malfunctions in medical devices. If we refer to developing countries, here the problem is the limited financial resources. In this case, a proper management of medical technologies, based on increasing reliability and reducing failures, could lead to the provision of good health services in limited economic conditions. Assessing the effectiveness of any maintenance program is critical to optimizing the use of available resources within the hospital. The emphasis cannot be placed solely on scheduled maintenance [22].

Patients' access to an accurate diagnosis, effective treatment, or rehabilitation process with appropriate medical devices is related to the efficient management of the maintenance of medical devices. This process can increase the efficiency and productivity of medical technology resources, which is especially important when resources are limited (**Figure 3**) [23].

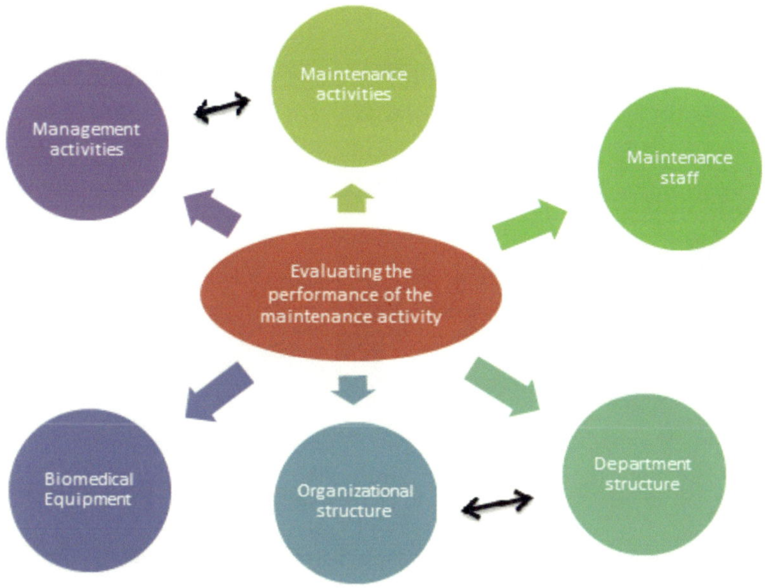

Figure 3.
A summary of basics for evaluating and identifying the mode of operation within the clinical engineering department of any hospital.

A possible approach to medical technology management involves the inclusion of a priority analysis step. The division into three levels of emergency (high, medium, and low) can be proposed based on the operation and implementation of subsequent actions in appropriate stages, appropriate to urgency, criticality, and seriousness. This analysis of priorities can be done both in terms of preventive maintenance, corrective maintenance, and in the case of a replacement program. In this case, the management priority can be recognized by the maintenance [24].

Another operational diagram that can be applied to evaluate the management process of medical technologies is the Ishikawa diagram or the fish bone diagram. This diagram is a verbal tool that has the relationship between an effect (a problem) and all possible causes that influence the effect (**Figure 4**).

The optimization of the functions of the maintenance department can be obtained by implementing a procedure that includes three important aspects: the task of the annual maintenance plan, the time available for the technical department, and the amount of equipment to be maintained. Taking into account these parameters, it is possible to estimate the personnel necessary to perform the maintenance function, responding to the most critical component of the management act – human resources [25]. Another approach involves designing a set of key performance indicators (KPIs) useful for assessing the performance of medical equipment maintenance in the management process (**Figure 5**).

A proper evaluation of the medical equipment maintenance process should include the development of a checklist for evaluating medical technology management. Implementing a list can be helpful in ensuring the profitability of health facilities and the reliability of medical equipment. In addition, she/he is involved in decision-making in support of the selection, measurement, repair, and maintenance of medical equipment, in particular for capital equipment managers and hospital medical engineers, and also for the evaluation of this process.

Figure 4.
Ishikawa diagram or the fish bone diagram applied to the management process of medical technologies.

An example of a maintenance management evaluation checklist proposes the inclusion of 15 indicators: type of medical equipment, quality control tests, application of training processes, correct storage of medical equipment and spare parts, existence of maintenance contracts, supervision of the process of operation, the existence of the decommissioning procedure, the existence of the reintegration system in use/system and the reporting of adverse events, ensuring the supply of electricity in optimal conditions, implementation of a process of continuous development of maintenance, general administration, management, and allocation of a separate budget for medical equipment maintenance services [23].

Another approach to evaluating the management process was implemented by Herrera-Galán and included evaluating the performance of the maintenance function by implementing management audits. Assessments include equipment availability, response to a service request, monitoring and control of biomedical equipment, staff training, quality of work performed by maintenance technicians, workload of maintenance technicians, control of work performed by maintenance technicians, effectiveness of annual planning, and maintenance and performance of the department. The results of this research show that the audit technique is a valuable checklist in evaluating the performance of a hospital. And in this case, it was highlighted that the most critical component of the results of a management audit is human resources [26].

Amerion et al. managed to identify the effective factors that influence the management of medical technologies at the level of a military hospital. Following the study, 26 components with an important influence on the management process were extracted. These are user training, human resources, user engagement and experience, the foreign exchange market, regular checks, and trade name. Attention to the evaluation of these components could reduce maintenance costs and increase the lifespan of medical equipment. The process of training users that must be continuous and the quality of human resources are the two main aspects [27].

A general conclusion regarding the implementation of management strategies in the maintenance of medical equipment shows us that the necessary adequate

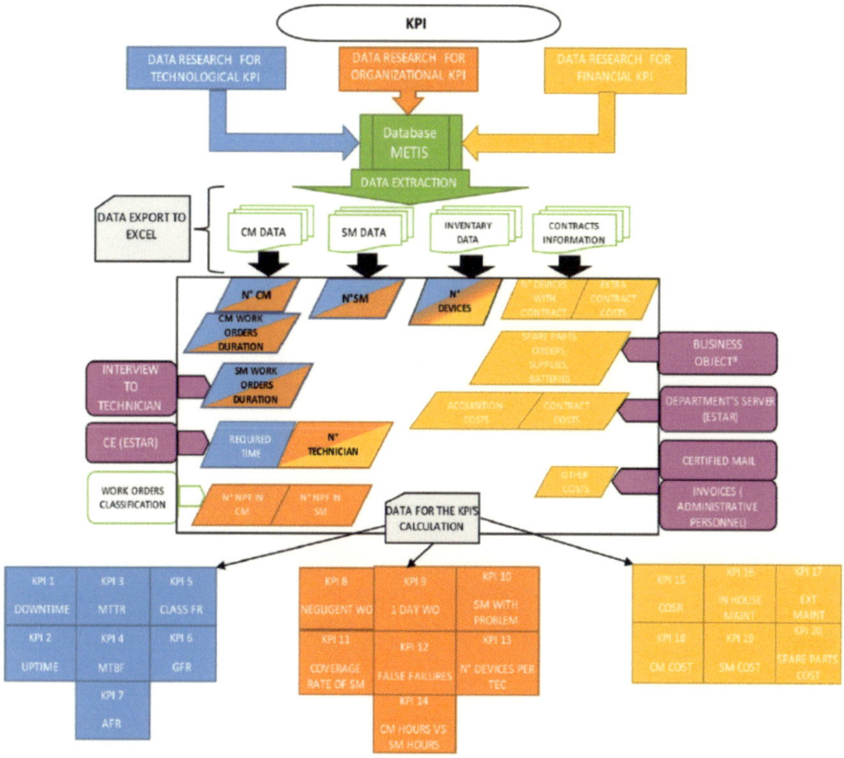

Figure 5.
Block diagram of the process required to calculate the set proposed by the KPI, starting from technological, organizational, and financial data.

resources underlying are human resources, material, financial, and documentation resources [28].

7. Conclusion

The maintenance of medical equipment becomes more expensive every year, and to optimize maintenance programs and reduce total cost of ownership, hospital management structures are constantly looking for solutions to extend the time of operation of equipment, in the required safety and technical performance and through the efficient use of available resources. The analysis and substantiation of capital expenditures in medical organizations must be based on quantifiable factors, with a direct impact on the full associated costs. In order to optimize operating and support costs, medical technology management structures develop medical equipment maintenance programs, based on assessments and prioritization, based on risks and costs. The development of alternative plans for medical equipment must take into account the complexity and large number of existing equipment, the skills of its own specialists and their number, the technical means of calibration and control, and the budgetary resources available.

In this regard, healthcare facilities need to implement evidence-based maintenance strategies, through the development of prioritization procedures aimed at a balanced assessment of relevant factors in the life of medical equipment, through an integrated approach to the elements of reliability-based maintenance, on risk-based conditions and maintenance.

Conflict of interest

The authors declare no conflict of interest.

Author details

Călin Corciovă*, Robert Fuior, Doru Andrițoi and Cătălina Luca
Medical Bioengineering Faculty, University of Medicine and Pharmacy Grigore T. Popa, Iași, Romania

*Address all correspondence to: calin.corciova@umfiasi.ro

References

[1] Kihiu J, Maranga S, Mutia D. Maintenance management of medical equipment in hospitals. Industrial Engineering Letters. 2012;**2**(3):9-19

[2] Patil PJ, Patil SP, Jaltade VG, Gupta SS. Departmental equipment maintenance system in government medical college. International Archives of Integrated Medicine. 2015;**2**(3):79-86

[3] Bahreini R, Doshmangir L, Imani A. Affecting medical equipment maintenance management: A systematic review. Journal of Clinical and Diagnostic Research. 2018;**12**(4):1-7

[4] Bahreini R, Doshmangir L, Imani A. Influential factors on medical equipment maintenance management. Journal of Quality in Maintenance Engineering. 2019;**25**(1):128-143

[5] Subhan A. Equipment Maintenance, Biomedical. In: Webster JG, editor. Encyclopedia of Medical Devices and Instrumentation. NJ USA: John Wiley & Sons, Inc.; 2006. pp. 289-321

[6] Dyro JF. Introduction to the Clinical Engineering Handbook. USA: Elsevier Academic Press; 2004. ISBN: 9780080476575

[7] Hegarty FJ, Amoore J, Scott R, Blackett P, McCarthy JP. Chapter 7 - the role of clinical engineers in hospitals. In: Clinical Engineering. USA: Elsevier Academic Press; 2014

[8] Corciovă C, Andrițoi D, Luca C, Ciorap R. Prioritization of medical devices for maintenance decisions. In: Vlad S, Roman N, editors. International Conference on Advancements of Medicine and Health Care through Technology; 12th - 15th October 2016, Cluj-Napoca, Romania. IFMBE Proceedings. Vol. 59. Cham: Springer; 2017. DOI: 10.1007/978-3-319-52875-5_67

[9] Andrițoi D, Luca C, Corciovă C, Ciorap R. Predictive maintenance application for health technology management. 2013 8th International Symposium on Advanced Topics in Electrical Engineering (ATEE). IEEE; 2013. pp. 1-4. DOI: 10.1109/ATEE.2013.6563431

[10] Corciovă C, Andritoi D, Ciorap R. Elements of risk assessment in medical equipment. In: 2013 8th International Symposium on Advanced Topics in Electrical Engineering (ATEE). IEEE; 2013. pp. 1-4. DOI: 10.1109/ATEE.2013.6563427

[11] Brown AS. Chapter 6 - Risk Management. In: Taktak A, Ganney P, Long D, White P, editors. Clinical Engineering. USA: Elsevier Academic Press; 2014. pp. 77-92. ISBN: 9780123969613. DOI: 10.1016/B978-0-12-396961-3.00006-8

[12] Anderson FA. Medical device risk assessment. In: The Medical Device Industry: Science, Technology, and Regulation in a Competitive Environment. New York: Marcel Dekker Ltd.; 1990. pp. 487-493

[13] Goode J. Use of international standard ISO 10993-1. In: Biological Evaluation of Medical Devices - Part 1: Evaluation and Testing within a Risk Management Process. Food and Drug Administration, HHS. 2016

[14] Cox LJ. What's wrong with risk matrices? Risk Analysis. 2008;**28**(2): 497-512

[15] AS/NZS 4360:2004. Risk Management Standard

[16] Amoore JN, Ingram P. Learning from adverse incidents involving medical devices. British Medical Journal. 2002;**325**(7358):272-275

[17] ISO 14971:2007. Medical Devices - Application of Risk Management to Medical Devices

[18] ISO 31000:2009. Risk Management - Principles and Guidelines

[19] Sezdi M, Two different maintenance strategies in the hospital environment: Preventive maintenance for older technology devices and predictive maintenance for newer high-tech devices. Hindawi Publishing Corporation. Journal of Healthcare Engineering. 2016;**2016:**7267983. DOI: 10.1155/2016/7267983

[20] Medical Equipment Maintenance Programme Overview. WHO Medical Device Technical Series, Department of Essential Health Technologies World Health Organization, Switzerland. ISBN: 978 92 4 150153

[21] Iadanza E, Gonnelli V, Satta F. Monica Gherardelli1, evidence-based medical equipment management: A convenient implementation. Medical & Biological Engineering & Computing. 2019;**57**:2215-2230. DOI: 10.1007/s11517-019-02021-x

[22] Kihiu J, Maranga S, Mutia D. Maintenance management of medical equipment in hospitals. Industrial Engineering Letters. 2012;**2**(3):9-19

[23] Arab-Zozani M, Imani A, Doshmangir L, Dalal K, Bahreini R. Assessment of medical equipment maintenance management: Proposed checklist using Iranian experience. Biomedical Engineering Online. 2021;**20**:49. DOI: 10.1186/s12938-021-00885-5

[24] Zamzam AH, Al-Ani AKI, Wahab AKA, Lai KW, Satapathy SC, Khalil A, et al. Prioritisation assessment and robust predictive system for medical equipment: A comprehensive strategic maintenance management. Frontiers in Public Health. 2021;**9**:782203. DOI: 10.3389/fpubh.2021.782203

[25] Herrera-Galán M, Martínez-Delgado E. Management audit applied to the maintenance department in hospital facilities. Ingeniería Mecánica. 2017;**20**(3):152-159ISSN 1815-5944

[26] Herrera-Galán M. Management audit applied to the maintenance department in hospital facilities. Ingeniería Mecánica. 2017;**20**(3):152-159

[27] Amerion A, Alijanzadeh M, Teymourzadeh E. Effective factors on the management of medical equipment maintenance in a military hospital: A qualitative study in Iran. EBNESINA. 2015;**17**(3):11-18

[28] Wang B. Medical equipment maintenance: Management and oversight. Synthesis Lectures on Biomedical Engineering. 2012;**7**(2):1-85